STRATEGIES FOR EXPLOITATION OF MINERAL RESOURCES IN DEVELOPING COUNTRIES

Proceedings of the International Symposium organised by the Indian School of Mines, Dhanbad, from 6 to 8 November 1986 on the occasion of its Diamond Jubilee.

Edited by
AJOY K. GHOSE

INDIAN SCHOOL OF MINES

A.A. BALKEMA/ROTTERDAM/1987

INDIAN EDITION SERIES

1. P. Ramakrishnan (Editor): Powder Metallurgy Alloys.
2. P.K. Jain: Major Carbon Industries in India.
3. S.K. Ramiah and L.S. Chickanagappa: Soil Mechanics and Foundation Engineering.
4. R.K. Sinha: Industrial Minerals.
5. T.N. Ananthakrishnan: Bioresources Ecology.
6. CMERI: Handbook of Open-Web Structures.
7. N.S. Subba Rao: Biofertilizers in Agriculture.
8. C.M. Srivastava (Editor): Recent Advances in Materials Research.
9. Cecil J. Saldanha: Flora of Karnataka.
10. P.K. Das & S.L. Srimani: Handbook for the Design of Castellated Beams.
11. H.C. Gangulee: Handbook of Indian Mosses.
12. Mary-Frances Thompson, R. Sarojini and R. Nagabhushanam (Editors): Biology of Benthic Marine Organisms.
13. S.N. Ghosh: Flood Control and Drainage Engineering.
14. K.K. Trivedi (Editor): Fisheries Development: 2000 A.D.
15. S.K. Jain: Ore Processing.
16. A.K. Ghose (Editor): Strategies for Exploitation of Mineral Resources in Developing Countries.

INDIAN EDITION SERIES 16

ISBN 90 6191 498 1

PRINTED IN INDIA

FOREWORD

The widening gap between the North and the South has become a central issue of our time. Since mineral development, as a part of industrialisation, plays a critical role in the process of economic development, many developing countries with mineral endowment have followed this route purposefully and with determination. In the context of the current regressive changes in the international mineral industry scene with depressed markets and prices, the future of mineral development programmes in many LDCs is seriously threatened. Besides structural changes in the world economy and a new order of multilateral cooperation, developing countries have also to adopt pragmatic approaches in their mineral policies and strive for the new decade of development.

This volume collates 23 papers on the varied facets of the problems of mineral development in LDCs, ranging from the role of government to human resources development, from the issues of environment to small-scale mining and its contributions, that would form the basis of a major dialogue at the International Symposium being hosted at Indian School of Mines, Dhanbad from November 6 to 8, 1986. The Symposium, organized as a part of the Diamond Jubilee activities of this prestigious mineral industry institution, should help stimulate awareness of the problems of mineral development in LDCs and motivate a search for effective solutions, both nationally and internationally. In an increasingly interdependent world, the burgeoning mineral needs of the global population can only be met if we treat our mineral endowment as a common heritage of mankind and work towards common goals with a spirit of accommodation. The Symposium, hopefully, will provide a forum for exploring the road maps for mineral development in developing countries.

I should like to express sincere thanks to the Organizing Committee and to the authors for making it possible to bring out this Symposium volume and to the co-sponsors for their generous support.

Indian School of Mines, Dhanbad
September 1986

AJOY K. GHOSE
Professor of Mining Engineering &
Symposium Convenor

ORGANIZING COMMITTEE

Professor D.K. Sinha	Chairman
Professor A.K. Ghose	Convenor
Professor T.C. Rao	
Professor B.B. Bhattacharya	
Professor Dinesh Chandra	
Professor M. Ramakrishna	
Professor H.M. Kaul	
Professor K.L. Rai	
Dr. P.K. Gangopadhyay	
Shri S. Laik	

CONTENTS

LIST OF AUTHORS

A.R. Adusumilli, Special Communication Specialist, SQN 210, B1.A., Brasilia, Brazil.
S.V. Ali, Mining Advisor, Mineral Development Board, 11 Tolstoy Marg, New Delhi, 110 001, India.
T. Atkinson, Professor & Head, Department of Mining Engineering, The University, Nottingham, NG 7 2RD, England.
S.P. Banerjee, Professor & Head of Mining Engineering, Indian School of Mines, Dhanbad, 826 004, India.
D.N. Bhargava, Controller General, Indian Bureau of Mines, Nagpur, 440 001, India.
A. Bhaskara Rao, Professor, Department of Geosciences, University of Brasilia, 70910 Brasilia, Brazil.
B.B. Bhattacharya, Professor & Head, Department of Applied Geophysics, Indian School of Mines, Dhanbad, 826 004, India.
S.A. Bilgrami, Chief Executive, Resource Development Corporation (Pvt) Ltd., Sheikh Sultan Trust Building, Beaumont Road, Karachi, Pakistan.
A.N. Bose, Regional Controller of Mines, Indian Bureau of Mines, Hyderabad, India.
Claudio de Castro, Dept. de Engenharia de Minas e CECINE, Federal University of Pernambuco, 50,000 Recife, Brazil.
B.B. Dhar, Professor & Head, Department of Mining Engineering, Benares Hindu University, Varanasi, 221 005, India.
A.K. Ghose, Professor of Mining Engineering, Indian School of Mines, Dhanbad, 826 004, India.
C.S. Jha, Managing Director, Bihar State Mineral Development Corporation, Raj Hotel Building, Main Road, Ranchi, 834 001, India.
M.B. Katz, Professor, Department of Applied Geology, School of Mines, University of New South Wales, P.O. Box 1, Kensington, N.S.W. Australia.
G.S. Khuntia, Asst. General Manager (MEQ), Rourkela Steel Plant, Rourkela, 769 011, India.
T.B. Malhotra, Consulting Mining Engineer, B-180, East of Kailash, New Delhi, 110 065, India.
J.F. McDivitt, Indonesian Mining Association, Jin Sutan Syahrir No. 6, Jakarta, Indonesia.
S.K. Mukhopadhyay, Assistant Professor, Department of Mining Engineering, Indian Institute of Technology, Kharagpur, 721 302, India.
Akhileshwar Singh, Deputy Advisor Environment, Department of Forest and Environment (Bihar), Patna, India.
K.L. Rai, Indian School of Mines, Dhanbad.
S.K. Sarkar, Central Mining Research Station, Dhanbad.
Ann Seidman, Brandeis University, USA.
Balbir Singh, Senior Deputy General Manager, Hindustan Copper Limited, Calcutta, 700 017, India.
R.D. Singh, Visiting Professor, Department of Mining Engineering, Kothagudem School of Mines (Osmania University), Kothagudem, India.
D.K. Sinha, Director, Indian School of Mines, Dhanbad.
S. Srini-Vasan, Indian School of Mines, Dhanbad.
Rajeshwar Thakore, Director, Aryan Mining & Trading Corporation Ltd., Calcutta, India.
R.K. Verma, Professor, Department of Applied Geophysics, Indian School of Mines, Dhanbad, 826 004, India.

KEYNOTE ADDRESSES

MINERAL DEVELOPMENT STRATEGIES IN THE DEVELOPING COUNTRIES

T Atkinson
University of Nottingham, U.K.

INTRODUCTION

Mining investment has traditionally been financed through (a) the internal cash flows of mining corporations or (b) through issues of new equity. The international loan institutions have also provided considerable capital especially in the heyday of international mining investment after the second world war. More recently, rapid rising energy costs; delayed reaction to the wave of expropriations and nationalizations that took place in the late 1960's and early 1970's; world-wide inflationary pressures; massive infrastructure and project installation costs; increasing intervention by the governments of both the developed and the less developed countries (LDCs); and the recessionary state of the world economy have restricted the flow of much needed high-risk investment capital to the world's mineral industry. The mineral industry is substantially failing to provide the level of investment needed to secure future mineral supplies.

Multinational corporations, after their recent experiences in the developed countries and especially in the LDCs, now adopt a very cautious approach to international mineral investments. Over the past 20 years, various UN agencies have attempted, with some success to increase mineral exploration and development activities in the LDCs, but their efforts are inhibited by lack of positive response from the developed world's mineral community and governments (Carman, 1979).

While the LDCs, the probable source of most future mineral supplies (Butt, 1982), are beset with an acute lack of mineral development finance, the present policies of governments in the developed countries are close to economic introversion. There are hardly any indications to suggest that mineral investments are taking place at the required level. It is however in the interest of all to pave the way for increased investments in the development of mineral resources wherever they occur to ensure adequate future supplies of mineral raw materials at acceptable prices.

According to a study initiated by the British-North American Committee (BNAC) (BNAC. New York, London, 1976) and a later study by Radetzki and Zorn (Radetzki and Zorn, 1979), $12 billion/year (1977 dollars) will have to be invested in non-fuel mining and processing facilities to see us through to the year 2000 if no disruptions in the supply of minerals are contemplated. As is widely reported by a number of studies (Bosson and Varon, 1977) LDC sources of mineral production may already account for 40-50% of present mineral supplies, and it is therefore a fairly safe assumption that out of that $12 billion at least $4.5 billion/year investment in the mineral industries of the LDCs is needed (with a proviso that economic growth in the coming years is not very high).

Huge shortfalls in mineral investment seem almost inevitable (Atkinson, 1979) and do not augur well for the future of the international economy, especially when seen in the light of long lead times generally required for exploration and developing an ore body into an operational mine. Furthermore, the days are gone when exploration budgets of, say, $500000 would suffice to initiate a sizeable mineral discovery programme, for, in addition to price rises, the easily accessible, exposed and near-surface mineral deposits near to centres of population have mainly been discovered and the search for deep, hidden, mineral deposits in often inhospitable terrains, is difficult and expensive. Exploration budgets are more likely to be of the order of several million dollars.

It is against this background of rising mineral exploration and development costs, long lead times, the inherently risky nature of mineral investments and the lacklustre performance of international mineral finance that the shortfalls in global mineral investment levels exhibit their most dangerous ramifications. Trends and attitudes developed over the last decade or so still continue and falling prices together with oversupply of most minerals hardly inspire investment confidence. "Although most metalliferous raw materials are now obviously in over supply there are many who recognize that the lack of sufficient capital investment in new mining projects is now sowing the seeds of serious future metal shortage"(Anon, 1975).

Slower growth in minerals consumption, increased recycling and development of substitutes could provide some relief to the shortage, but world demand will still require much larger investment than is currently being provided. While the mineral consuming countries may be content with the present oversupply of minerals and low mineral prices in the international markets, any complacency at this point could lead to a very uncertain future. It is only through sustained and substantial investments in all phases of the world mineral industry that mineral supplies can be ensured.

SOURCES OF MINERAL FINANCE

There is a multitude of sources of finance available to the world mineral industry. While collectively the likely finances available from these sources are far in excess of the needs of the mineral industry, each source has different criteria for releasing its funds, depending on its needs, interests and priorities.

Domestic sources of finance in LDCs

A primary feature of most LDCs is their lack of domestic capital. Private investors in these countries, because of the nature and size of mineral investments, are more interested in consumer product industries where profits are good, relatively low-risk and immediate. Most LDC governments, on the other hand, may want to undertake mineral development from their own meagre public funds, but find it hard to justify such expenditures in view of other highly pressing social and economic trends.

Some LDC governments have attempted to channel public funds through state-owned corporations for the development of mineral resources. Brazil, Zambia, Chile, Peru, Zaire, Iran, Bolivia, Burma, India and Indonesia have all used this method (Skelding, 1969). There are cases in which the host government has relied mostly on its own financial resources (Tilton, 1977) but most LDCs do not possess adequate financial resources to undertake major mining projects on their own. Their ability to borrow from international financial institutions is limited because of their already heavy external debt.

Conventional sources of finance

The Multinational corporations. Much of the high-risk exploration and the development effort has traditionally been funded either through their own cash flows or equity share and managed by multinationals from the USA, Canada and Europe, often with assistance of Japanese capital. These corporations have acquired over the past 50 years or so a virtual monopoly of the various skills and techniques required to develop and operate the mining industry. Although their financial resources have come under great strain because of the current international economic climate, they still have the ability and the capacity to invest in mineral resource developments wherever they occur. With ready access to their own capital and high credit rating with the international financial institutions the multinationals are able to spread their risk by carrying out a number of activities and operating in a number of regions, often in joint ventures. Although engineers, geologists and other specialists can easily be hired by governments or national mining companies, the fully integrated teams needed to develop large projects can rarely be assembled without the cooperation of the multinationals. With their own financial resources and a near-monopoly of techniques, methods and processes, the multinationals occupy a unique position, the implications of which cannot be ignored when discussing the question of financing international mineral developments.

Commercial banks. The developed countries' commercial banks, both national and international, are the biggest international source of private capital. In recent years the larger multinational banks have favoured participation with multinational corporations to invest in the mineral industry and have provided substantial loan capital to the multinationals, including that for mineral projects located in LDCs. However, the banks require their capital fully guaranteed against such risks as expropriation, inconvertability, war, revolution and insurrection. Undertakings to this effect by the host governments alone may not be adequate to placate their fears. The insistence of the commercial banks on a continuing role for technically qualified companies and their re-verification (Anon, 1978) of the economic viability of the mineral project before any project loan application is considered appear to confirm that a multinational must be a part of any mineral development programme, no matter where it is undertaken. Furthermore, the banks are reluctant to give loans to developing countries because of their mounting external debts and their potential inability to pay them back (Newsweek, 1981).

Bilateral and multilateral aids. In the past this kind of assistance has been provided to LDCs, particularly by Canada, France, the UK, the USA, FR Germany and the USSR. A number of other intergovernmental and regional organizations such as OECD, Colombo Plan, the Commonwealth, NATO, etc have also been active in providing technical assistance in the general area of mineral development. However, these services are short-term and advisory in nature rather than involving large-scale operational activities. While the LDCs have benefited from these aid programmes, many have questioned the intentions, mode, amount and effectiveness of such assistance (Atkinson, 1968). However, with a changed outlook and approach responsive to future mineral needs, the governments of the developed world can certainly provide a boost for mineral development activities in the LDCs.

Export credit institutions. Supply of equipment on credit, equipment leasing, merchant finance, etc. do provide some financial help, but this is only remedial in nature and generally available only for on-stream projects. The contribution of these sources of finance to international mining development is therefore limited.

Non-conventional sources of finance

In addition to traditional sources of mineral finance, oil companies, and oil-producing countries, and insurance companies have recently emerged on the international financial scene as providers of finance. With their large cash flows and their long working associations with LDC governments oil companies are well equipped to become involved in mineral projects. This involvement has taken the form of either setting up new mining subsidiaries and entering all phases of mining or outright acquisition of operating mines (Fells, 1978), and could provide some relief to the capital-starved mineral industry. The oil companies may be able to persuade the international financial institutions to provide them with loan capital. However, recent indications are that the oil companies are prepared to summarily cease their mining investments.

Although the current financial health of most OPEC members may be suspect, petrodollars have played a role in financing mineral developments. Much of the liquidity held in euro-currency is provided by these peltrodollars and the amount that is recycled for mineral development is not precisely known, but there have also been a few cases in which Arab states have made funds available for specific mining projects. So far, direct investments from the OPEC members have not been very impressive, but there is scope for channelling petrodollars either directly or indirectly into the development of the world's mineral resources.

In common with oil companies, insurance companies in the developed world are another major source of capital. Because of their stable and predictable financial structure, they are better suited to advancing long-term loans to mining corporations. These companies have long been active in the iron ore sector of the USA, but their involvement in other mineral projects, especially in LDCs, is quite recent. There are at least two major mining projects which have partly relied on insurance companies' financing; Erstberg Copper Project in Indonesia and the Falconbridge-Dominicana Nickel project in the Dominican Republic. These insurance companies have, so far, made their funds available to projects owned by the private sector. With their vast financial resources and capacity to play an important role in any mechanism of underwriting investments in LDCs, they are potentially useful partners in the development of mineral resources.

International agencies

The World Bank, either on its own or through its subsidiary bodies has provided up to $1,000 million in loans for assisting mineral-related developments in LDCs since 1957. It is clear that the World Bank is not a major contributor to international mineral finance, but it has often served as a catalyst for much larger loans and investment from other sources. It has also provided technical assistance in the early stages of project development and undertaken surveys that have led to identification of viable mining projects.

Although the efforts of the various UN agencies are commendable, they are hampered because of the lack of adequate funds at their disposal and because the necessary back-up from the world mineral community and the governments of developed countries is lacking.

CONSTRAINTS ON MINERAL INVESTMENTS

There are several reasons for the failure of the international mineral industry to invest adequate sums of capital. Uncertain economic conditions coupled with the inflationary pressures on the establishment of new mine and plant capacities have limited the industry's capacity for investments. High bank rates in the early 1980's not only caused cash flow problems for the industry, but further limited the availability of risk capital for mineral ventures. Cyclical market behaviour of most of the metals prices does little to inspire any confidence amont the investors. While some of these reasons are transient in nature, a major stumbling block to mineral investments, especially in the developed countries, springs from the general attitude of the people and their governments on issues such as environmental protection, its regulatory controls and the withdrawal of large areas of land in the name of plant and animal life and recreation.

However, the most important constraint has been the attitudes of host governments towards foreign investment in their mineral industry sector. Even countries like Canada and Australia have recently sought to increase their share and control of profits from the development of their mineral resources. While government pressure has at present been partially relaxed in Canada and Australia, there are no indications to suggest a similar change of attitude on the part of LDC governments claiming a bigger share of profits from mineral projects and insisting on 'indigenization' of their mineral resource industry. The problem is further accentuated by multinationals' lack of trust in the political and economic stability of the LDCs.

While the developed countries becoming increasingly dependent on the mineral resources of the LDCs, any financial equation for the development of the world's mineral resources must revolve around the multinational corporation and the LDC - the two most important denominators determining the pace and course of international mineral development. Absence of any mutually acceptable mineral development approach between the two will inevitably lead to an uncertain future for mineral raw material supplies - precisely the environment of the international mineral scene today.

CONFLICTS OF INTERESTS BETWEEN MULTINATIONALS AND LDCs

There is no doubt that 'the main change in the world mining industry this century has been the increasing dominance of the multinational corporations, coupled with rising demand for a whole range of new minerals' (Govett and Govett, 1976). The vast financial and technical resources accumulated through their operations all over the world, combined with easy access to international finance, place the multinationals in a unique position. Their dominance in the mineral industry is further strengthened both by integrating backward to obtain ore supplies and by integrating forward to the processing and/or fabricating stage of a particular mineral, thus extending the scope and range of their activities and dominating, if not controlling, not only production but also the major marketing institutions.

From this position of strength multinationals have certain distinct leverages that are not available to, say, a state-owned company. They extract resources from one country, manufacture products from these resources in a second, and sell the products in a third. Thus the geographical distribution of their activities enables them to capitalize on the special advantages of a particular country.

It is irrelevant whether the multinationals undertake mineral developments in developed or less developed countries, their objectives are geared to growth and maximization of their profits. Ideally speaking, the multinationals would like de facto control of mineral operations in the host country; complete or majority ownership; full management and operational control over production, employment, finances, investments, purchases, marketing and distribution; freedom to deal with their parent and/or subsidiary companies; little or no processing in the country of extraction unless justified by their overall strategy; favourable and stable fiscal regimes; a minimum of 30 years of tenure (covering all phases of mineral resource development) with financial legislative guarantees and compensation in case of expropriation; no re-negotiation of mineral investment agreement over the life of the contract, with a provision to take the host government to the international arbitration in case of a dispute (Miksell, 1980).

Such an approach is totally untenable in the present day LDC where resources are considered part of the national heritage, the development of which should not only be under national control and subject to maximization of profits locally, but the benefits of which should permeate the whole economy as well. On their part, the LDCs would like to see their sovereignty over their mineral resources restored; 100% local ownership; an effective voice in the management, policy making and control of the project regarding production, sales, marketing, distribution, etc; training of locals and replacement of expatriates at all levels; economic development and environmental protection provision; monitoring the foreign deals of the multinationals, especially those with their parent/subsidiary companies; local processing and foreward integration; application of national practices, laws and institutions on taxes, imports, exports and foreign exchange transactions of the project; separate contracts for exploration and development; re-negotiation when it is in the national interest; and dispute settlements in accordance with their national laws.

These then are the diametrically opposite viewpoints of the multinational corporations and the LDCs. Pre-independence practices and policies of foreign investors and their continued adherence to the same basis for investments in the present-day LDCs are unacceptable to LDCs. Had there been alternative choices available, mineral developments in the LDCs would have ceased to exist long ago. But, because of the needs and interests of multinationals (and their home governments) and the LDCs are almost inseparable, new mineral agreements will continue to be negotiated and signed, although not at the rate necessity demands.

Possible compromises

Mineral agreements signed in the 1970s (Bougainville, Papua New Guinea, 1974; OK Teddi, Papua New Guinea, 1976; Corro Colarado, Panama, 1976; Teluk Tomini, Indonesia, 1977; and El Indio, Chile, 1977) shed some light on possible compromises and to some extent accomodation of the opposite viewpoint. The multinationals have accepted, in principle, the sovereignty of the host governments over their mineral resources; and share in equity ultimately leading to 100% local ownership (on a pre-agreed formula); some voice in the managment and control of the project; hiring and training of locals; economic development provision; and some role for the federal bank in the foreign transactions of the project. The LDCs have agreed to accelerated depreciation and amortization (lower take for the LDC government in earlier years of the project); a sliding scale of corporate tax based on discounted cash flow rate of return (DCFROR); international arbitration; and highly-geared external finance (tacit approval for the disguised equity of the multinational).

The present US administration and UK government feel that the best solution to North-South disparity is through private investment, obviously from the multinationals. These governments regard inter-national institutions, including the UN and the World Bank,as global equivalents of the welfare state undermining the principles of self-help and free enterprise (Sampson, 1981). Business corporations are being urged to search deeper into their hearts and dig deeper into their pockets to improve the lot of the people. The Reagan administration's plea to the multinationals to lead the development in the less developed world is not new; the great trading corporations of the 18th and 19th centuries proclaimed the same worthy intentions. Nevertheless, these corporations were and still are in the business to make money and not to develop LDCs except as markets or sources of raw materials (Vittachi, 1982).

The World Bank is by far the most influential international financial institution. Its role as a lender could be substantially extended if adequate back-up is available from other groups of the world mineral industry. Sustained World Bank Group involvement through the acquisition of equity; and the negotiating, planning, financing and technical re-appraisal facets of the project could provide a powerful balancing effect in reconciling the differences of investors and the host governments. Such a proposal was sponsored by the UN and discussed in Washington by participants from major multinationals, public servants concerned with world-wide mineral developments and World Bank officials. The Bank's attempts to increase its participation in financing and developing the world's mineral resources was heartily endorsed by all the participants, although what became of the recommendations of this meeting remains unclear.

If the current minerals shortfall persists, some form of involvement by the World Bank or another world body that oversees mineral developments, especially in the LDCs, and keeps watch over the interests of all those involved may ultimately become inevitable. Such an approach may be in the offing for seabed minerals, where the UN is likely to play a pivotal role (Kimball, 1982). A similar UN involvement in areas which supply the bulk of future mineral supplies would seem appropriate, but it must be remembered that mining does not seem to thrive in a political/legalistic environment and that history has shown that private sector operations are invariably more efficient and can provide a better cost-benefit ratio.

THE HUMAN ASPECTS

So far this address has been concerned with attracting finance from the loan institutions. Mining developments are however concerned with people as much as with money. There is a chronic world shortage of engineer/administrators with international experience of major mining developments. India however has more professional engineers than any country in the world, but is it any better off than other countries ? According to a Japanese management consultant the following applies (Kharbanda, 1983):

1	Indian engineer	=	2 Japanese engineers
2	Indian engineers	=	1 Japanese engineer
3	Indian engineers	=	Utter confusion

Why should this be so ? The Indian nation has produced a far greater share of international entrepeneurs who have gone out into the world, built large corporations and created wealth for their host countries and for themselves.

Undoubtedly team work produces project success. There are people who can work together in a team, who meet informally in small, work oriented groups, that can produce the results that matter. The large meetings of the "heavy weights" with reputations to protect may not be so concerned with progress. A slightly cynical but unfortunately true picture of progress in a mining project is (Atkinson, 1986):

1. Enthusiasm
2. Disillusionment
3. Panic
4. Search for the guilty
5. Punishment of the innocent
6. Honours for the non-participants

Creating the right environment

It appears essential then that an environment where initiative and progress will thrive must be created if mineral developments are to be successful. How can this be done in the short term ?

South Korea lost almost the whole of its industrial base in the war of the 50's. Its young male population was decimated, but between 1973 and 1978, starting from scratch, Korean construction contractors earned more than three times the volume of work earned by British contractors in the Middle East. South Korea is one of the few countries with a positive trade balance with the OPEC nations (Atkinson, 1986)

This has not come about by accident. The Koreans have created their own entrepeneural environment:

- Korean engineers are more flexible, think laterally and are not bound by the traditional engineering disciplines. They are more adaptable than their Western counter-parts.

- Koreans are far more cost-conscious than other engineers and they meet target dates.

Korean craftsmen are better trained, are more highly skilled and usually have more than one skill, e.g. a fitter can double as an electrician, etc.

Koreans work more consistently and most important of all, work as a team.

All this has occurred in a country where tradition and introversion reigned supreme for four thousand years.

Almost certainly it is essential to be rid of stulifying bureaucracy. Control of projects from a distant head office does not work. It did not work in Roman times. When Hadrian's wall was built in Britain, it was decided in Rome that gateways should be provided at mile distant intervals. The wall was an administrative boundary to control movement between England and Scotland, not a fortification. At some mile gateways there are precipitous drops making the gateways useless. Very often rigid adherence to head office directives results in problems. Such procedures must be queried if the environment needed to make initiative and progress thrive is to be created. Those concerned must ask themselves is a procedure necessary? What does it contribute to the efficiency and economy of a project? Is a design over-elaborate? Is an item of equipment necessary? Years ago in the British coal industry we used the saying "If in doubt leave it out". We should develop mental flexibility again and forget the straight-jacket of inappropriate official procedures.

SMALL SCALE MINING

For decades the international mining industry has been fascinated by large scale operations which can achieve so-called "economies of scale". Such projects require huge financial investments beyond the resources of the LDCs. Larger scale can be cheaper in terms of capital expenditure per annual tonne of production, but the apparent initial economies of scale can be offset by the many additional problems that large scale brings with it:

- a more complex organisation
- reliability is more suspect
- commissioning is more difficult
- operating and maintenance personnel must have greater skills and knowledge
- disruption of supplies to the plant is costly
- because construction takes longer there are greater opportunity for delays
- there is increased environmental disturbance

In the LDCs economies of scale are not always achievable and "small may be beautiful".

The advantages of small scale mining

These have been comprehensively described elsewhere (Wells, 1984) and may be briefly listed as:

1. Relatively short lead time.
2. Rapid conversion of resources into benefits.

3. Greater exploration success.
4. Usually indigenous capital is used and indigenous control of operations is achieved.
5. Local "downstream" industries usually form the market, providing the equivalent of local forward integration, which is absent in many large projects.
6. Profits are generally retained in the country.
7. They greatly benefit the local economy, they have a "multiplier effect" and provide an improved standard of living.
8. Small-scale mining responds more quickly to mineral price changes and provides greater price stability.
9. There are benefits in the training of managerial, technical and craft skills. Those trained in small scale mines are generally more self-reliant and versatile.

These advantages are so important that there is a strong case for setting up a small scale mining bureau using "mining uncles" (Wells, 1984) to provide technical assistance. Unfortunately modern industry has failed to recognise the importance of small scale mining and there are few mining engineers available today with the experience and breadth of skills to act as mining uncles. It is also an area neglected by the educational institutions and courses in the intermediate technologies appropraite to small scale mining should be organised, and made available to the mining engineers of the developing nations.

Small scale mining requires special people with a combination of resilience, determination, versatility, resourcefulness, cheerful optimism and above all an entrepeneural flair. Such people are rare but should be nurtured, encouraged and given the opportunity to bring prosperity to their locality.

CONCLUSIONS

Many will say that in a time of recession the developing nations should not be concerned with new mining activities. Many mining corporations have metal inventories as large as 18 months normal production; prolonged mine closures and minimal sales have catastrophically reduced their cash flows and even the previously top performers are repeatedly reporting losses with little hope of returning to profits. Exploration efforts are at an all time low. Geologists, drillers and other exploration personnel have been laid off in unprecedented numbers; but the long lead times associated with mineral projects mean that we should be planning now to meet world mineral demand when we climb out of the present trough of depression.

It is doubtful whether the huge mining complexes of the past, e.g. North West Australian iron ore, etc. will be repeated in the future. The developing nations will be far better placed to take a major part in such projects, providing them the opportunity for forward integration within their own countries, thereby giving them a larger share of the prosperity "cake".

The key to the situation rests with the leaders of the developed world, who must face the real issues of ensuring adequate mineral supplies while re-dressing the present imbalance in the sharing of the wealth created by mining minerals. There are many things the LDCs can do to help themselves but the major issue is the sustained effort needed by the developed nations and the financial institutions to reverse the trends set by the traditional pattern of mineral finance and development.

India with its huge domestic market, its engineering skills and its mining traditions is clearly in a very favourable position to become a major force in the world mining industry. It is an opportunity that must not be lost.

REFERENCES

1. Carman, J.S. Obstacles to Mineral Development - A Pragmatic View. Pergamon Press. Oxford, 1979.

2. Butt, N.A. Future Mineral Supplies and the Lesser Developed Countries. IMM Trans/Sect. A, Vol. 91, Jan 1982, pp A10-A17.

3. British-North American Committee. Mines Development in the Eighties: Prospects and Problems. BNAC. New York, London, 1976.

4. Radetzki, M. and Zorn, S. Financing Projects in Developing Countries. Mining Journal Books, London 1979.

5. Bosson, R. and Varon, B. The Mining Industry and the Developing Countries. Oxford University Press, Oxford 1977.

6. Atkinson, T. Future Strategic Mineral Supplies and the EEC. The Mining Engineer. No. 211, April, 1979, pp 721-728.

7. Anon. Another warning on metals famine to come. Mining Journal, Vol. 285, No. 7318, 1975, p. 385.

8. Skelding, F.H. Financing Mineral Developments in Under-developed Economies. Proc. Council of Economics of AIME. Feb. 1969, pp 33-46.

9. Tilton, J.E. The Future of Non-Fuel Minerals. The Brookings Institution. Washington, D.C. 1977.

10. Anon. What Bankers look for the Project Loan Applications. Mining Engineering, Dec 1978, pp 1646-1648.

11. Newsweek. 16th November, 1981.

12. Atkinson, T. Evaluating Mining Projects in Developing Countries. DIC thesis. Imperial College, London, 1968.

13. Fells, F.D. Financing New Mining Projects: Recent Trends and Prospects. 11th Commonwealth Mining and Metallurgical Congress, Hong Kong, May 1978.

14. Govett, G.J.S. and Govett, M.H. World Mineral Supplies - Assessment and Perspective. Elsevier Scientific Publishing Co. Amsterdam, 1976.

15. Miksell, R.F. Mining Agreements and Conflict Resolutions. In Mining for Development in the Third World. Pergamon Press, Oxford, 1980, pp. 198-203.

16. Sampson, A. A muted dialogue. Newsweek. 5th January, 1981. pp 50-51.

17. Vittachi, V.T. Reagan and the Third World. Newsweek. 8th Feb, 1982. p. 4.

18. Kimball, L. The Long Road Ahead: Third UN Conference on the Law of the Sea. 11th Session. Marine Policy. Vol. 6. No. 4. Oct 1982. pp. 342-344.

19. Kharbanda, O.P. and Stallworthy, E.A. How to Learn from Project Disasters. Gower, 1983.

20. Atkinson, T. The Challenge of World Mining. The Mining Engineer, September 1986.

21. Wells, T. Optimal Exploitation by Small Scale Mining. 12th World Mining Congress, New Delhi, India. Nov 12-23, 1984, pp 201.1 - 201.13.

STRATEGIES FOR STIMULATING NEW MINERAL DEVELOPMENT - THE ROLE OF REGIONAL COOPERATION

James F McDivitt
Indonesian Mining Association
Jakarta, Indonesia

ABSTRACT

In the rapidly changing mineral development scene new and different strategies must be identified, tested and adopted to help cope with new circumstances. In developing countries which often are limited in skilled manpower and financial support, regional cooperation is one approach which can bring benefits to participating countries.

Although regional cooperation can be on many different levels, it is most effective in areas of general impact such as training, research, information and to some extent policy and planning. Examples of regional programs in these areas are introduced.

One institutional framework for regional cooperation is the regional network linking institutions in several countries. A number of examples will be discussed including promotion of small-scale mining, environmental impact and development of regional information systems.

INTRODUCTION

Since before the dawn of recorded history man has made use of minerals as essential tools in his development. Over the centuries we have passed through successive ages -- stone, copper, bronze, iron, steel, nuclear -- and the evolution continues towards ever more complex systems at an ever more rapid pace. Today's mineral industry and thus its mineral strategy is remarkably different from that of even 20 years ago, and there is every reason to believe that the changes in the next 20 years will be at least as great. Changes on all levels -- the search for minerals and the minerals that we search for, the way that ores are extracted from the earth and broken down into their mineral components, the nature of markets and the trade patterns of minerals raw materials and products, the role of finance and the ways financial packages are put together, and perhaps above all the technologies which both allow and force the other changes.

But in it all there are a number of old-fashioned truths about minerals which do not change, which serve as under-pinnings of the economics of mineral supply, setting this field at the side of and slightly in conflict with classic economic theory. Though many aspects may change, minerals themselves continue to be fixed materials, fixed in location, fixed in size and fixed in physical and chemical properties including grade. Thus the basic stock in the earth must serve for all time, and will ultimately be exhausted. When a country has them mineral resources become a part of the natural heritage which can make an important contribution to development, and which must be used for the optimum long-term benefit of the people of the country. In stating this it should be noted that although deposits are fixed, the economic conditions under which any deposit can be developed very greatly, and that although minerals are a wasting or exhaustible resource there is no likelihood of a world-wide shortage of minerals.

Thus, when we talk of strategies for mineral development we talk of many strategies, since each country and each component of the industry has its own strategy, and each of these will be somewhat different, and often in conflict with others. Within a forum such as this, a degree of priority must be given to strategies designed to optimize the contribution of minerals to world well being, and by inference to the well being of the mass of people who make up the world. This implies, and even demands, cooperation among individuals, among districts within a country, and among countries, and it is this element of regional cooperation which will be emphasized here.

NATURE OF REGIONAL COOPERATION

In introducing this subject there are a number of points which should be taken into consideration:

- a region can be usefully thought of as a grouping of countries which have certain geological or physical features in common so that they can benefit from working together, e.g. the island arc countries around the Pacific Rim, or countries with a humid tropic or desert environment, or tin producing countries

- in practice, regional grouping should not be large -- a good range is between 5 and 10

- since mining is a complex and competitive business there are only certain aspects of mineral development in which cooperation is likely to be effective. The closer we are to the mine face the less likely we are to talk of cooperation, but this still leaves plenty of open ground

- any program of regional cooperation must have some form of institutional base to initiate and oversee the program

Within these limitations what are the areas in which regional cooperation can make a good and positive contribution to mineral development, and what are the best mechanisms or approaches to accomplish this?

In any exercise involving a number of countries, a good place to start is in non-controversial, non-confrontational areas, and good mechanisms and approaches are those which are highly flexible, controlled within the region, and have minimal costs and measurable benefits to each participating country. So far as areas for regional cooperation, four broad areas can be identified which qualify:

- training programs with a regional dimension
- cooperative research on problems of common interest
- development of information systems
- some areas of policy and planning

On mechanism or institutional framework, any type of regional linkage will serve, but one of the most flexible is the regional network approach. Here considerations will be given to these four areas and how they can be dealt with, but first a brief note on what is meant by a regional network.

REGIONAL NETWORKS

One of the most powerful mechanisms for regional cooperation is the regional network structure. A regional network consists of national institutions or bodies from a number of countries in a specific discipline or area of research and study, which are linked together to facilitate communication and cooperation. There are a number of underlying principles in establishing such networks:

- a major objective is to strengthen national capabilities through regional cooperation

- all activities are based on existing institutions and facilities, so there is no institution building

- benefits are spread to all interested scientists in participating countries, and these scientists have a high degree of control over policies and programs
- as this would imply, network structures are loose and flexible, and serve as a framework for cooperation.

Regional networks already exist in several scientific and technical fields and there is a good base of experience upon which to draw. A wide range of activities can be promoted within the network structure including training, seminars and conferences, cooperative research, exchange of personnel, publication of newsletters, etc. All this can be done within a relatively simple and inexpensive administrative structure, which usually is rotated from country to country on a periodic basis to share responsibilities, costs and benefits. Examples of existing or planned networks are included in the following sections, and a note on networks is attached as an Annex.

REGIONAL TRAINING

This is perhaps the most obvious area in which regional cooperation can make an impact, and there are already many good examples. A number of countries organize international training courses, for example, Japan has courses on offshore prospecting for minerals, and on geothermal energy. More relevant are courses given in other developing countries, and this is an area in which much useful work can be done.

The best example in Asia is the program of post-graduate traiing in geology sponsored by the Government of India. For several years now places have been reserved for candidates from the Asia/Pacific regional at the Geological Survey of India Training Institute in Hyderabad. The Institute has a 10 month orientation program to prepare young graduates to enter the national service in India. Each year about 100 candidates are given a varied program to familiarize then with the different aspects of geological survey work. This includes field work in different locales plus exposure to laboratory work, management and administration at the headquarters of the Institute.

In this case, the Government of India annually invites 10 participants from the ESCAP region, and covers all expenses within the country including subsistence, local travel, books, medical expenses, etc. Participants are given a thorough grounding in geological survey work.

Although most present training activities are based on bilateral linkages, regional networks are being established for training. One approach is to identify one or two training institutions in each participating country (ideally 8-10 institutions in all), each of which has existing expertise in one aspect of mineral development in which it can give specialized training for the region. This allows for some specialization, and a sharing of expertise, experience and expenses. It also helps build up a pattern of cooperation and promotes both institutional and individual linkages which go well beyond the specific programs in question. A cooperative program along these lines has been developed among engineering institutions in Africa, and the approach is followed by the United Nations University in some of its projects. Unesco or UNU would be likely imbrella organizations for such networks.

REGIONAL RESEARCH

Many developing countries have limited research facilities, and limited manpower and financial resources to devote to research projects. At the same time there are many problem areas which are somewhat unique to the developing regions e.g. exploration in areas of deep weathering, treatment of laterite ores, tailings disposal in areas of high rainfall, how to deal with small-scale mining and the informal mining sector, etc.

Although the critical mass of skills and money for broadly based research may not exist on the national level, it is allways true that research is being carried out, and one of the first tasks in any regional research program is to identify what is already going on. Regional linkages for research can provide a valuable service by doing no more than keeping a running inventory of ongoing research in the region and promoting exchange of information and personnel. This helps to develop communication and avoid duplication and leads to greater use of r esearch results. However, it is easy to go well beyond this, and to identify "regional centres of excellence" in particular branches of a subject, and activities which can serve as regional "pilot projects" for developing and testing research results. It is possible to put together a regional network involving a number of such pilot projects which complement each other, and which provide a much more complete coverage than any individual country could undertake on its own.

It is clear than any such research network would have strong training and information components, and would overlap into these areas, but since each overlapping network would have a somewhat different clientele this would be all to the good, as it would involve more specialists in the system.

REGIONAL INFORMATION SYSTEMS

The field of scientific and technical information is changing so rapidly that it is difficult to see the best path. There has been a good deal of discussion on the role of a regional mineral information network but this has moved slowly, perhaps for the best, as information technology has developed so rapidly that it is now relatively easy to establish the basic systems.

There is no doubt that much information exists which would be of value in the promotion of mineral development - information in reports, files, general literature and in a wide variety of sources, but it can only be of value if it is available and accessible. A regional mineral information system could be a big step forward through providing a framework for national systems, the links between them, and the training of information specialists to operate them, and mineral specialists to use them. A preliminary outline of such a project prepared at RMRDC envisaged four phases:

- a country by country assessment of existing needs, resources and facilities

- a regional meeting to identify the components of the system and outline its structure and operations

- a series of pilot projects on the national level to develop formats, division of responsibilities, etc.

- establishment of a regional network with emphasis on training of specialists and users, and development of flexible and compatible systems.

On a more specific level, a survey has been carried out on the possibility of establishing a network for coal information in Asia. Here it is clear that when detailed work is being done within a relatively narrow field there can be an important role for a specialized regional information network.

POLICY AND PLANNING

Policy and planning can be somewhat controversial on the national level. but tend to be less so regionally. In manycases it is possible to develop policy and planning guidelines regionally which are a help when translated to the national levels. Two recent examples will serve here. In November 1984 a workshop on mineral policy for small-scale mining was organized in New Delhi in conjunction with the World Mining Congress. This allowed national specialists and government officials to interact in a somewhat neutral setting, to identify areas in which changes are needed, and to explore ways of dealing with them. The second was a workshop on environmental impact of mineral development in the humid tropics, held in Manila in February 1985 which explored policy issues as they relate to environmental impact.

In both cases the workshops dealt with policy issues which are rather sensitive on the national level, but since they were regional workshop it was possible to discuss things in a more objective and detached way. At the same time the results can be used as a base for discussion on the national level. Here it is of interest to note that India has carried out an extensive national exercise on the place of small mines in the national mineral development system. As for the environmental impact workshop, it ended up with a set of guidelines for the environmental planning and management of surface mining in the humid tropics which can serve as a base for any country wishing to prepare such guidelines.

Similarly ESCAP has organized sessions on mining legislation in the region, and a recent meeting in Indonesia (July, 1986) considered the subject of Mineral Resource Assessment in National Planning and Policy Formulation. In the case of policy and planning, meetings may be a more flexible and acceptable mechanism than networks.

INSTITUTIONAL AND FINANCIAL ASPECTS

Regional projects must have an institutional base from which to operate, and in the case of larger projects should be sponsored and promoted by a credible body which can take responsibility for financial accounts. Regional networks as described above can serve in this function provided the host institution is willing to provide a degree of administrative support. There also are a number of regional institutions or organizations which operate regionally including UN bodies such as ESCAP, ASEAN, Asian Institute of Technology, etc., which can serve as an imbrella for regional programs.

Most of the costs of regional activities are already covered in the national budgets of participating countries, and in most cases it can be demonstrated that the regional dimension leads to savings and more efficient use of funds rather than the reverse. However, there are certain core expenses for communications, meetings, limited secretarial and administrative services, etc., which require additional funds. In many cases these can be absorbed by the host institution, but in other cases application can be made for small grants from technical assistance bodies, since it is possible to demonstrate that this small contribution has a significant multiplier effect in calling forth support from the region.

For simple cooperation projects a regional association or society can serve as the base provided it can supply limited core support. Where additional funds are to be generated it is normally necessary to have a more formal and permanent base to provide the necessary accountability.

CONCLUDING REMARKS

By encouraging regional cooperation in dealing with pressing problems related to mineral development we can accomplish a number of things:

- scarce manpower and financial resources are used more efficiently
- duplication is avoided by sharing research and development responsibilities
- results from a project can be tested and applied in other countries with similar characteristics
- projects can serve multiple functions of research, training and demonstration, both in the host country and in the region
- regional involvement highlights projects in each of the participating countries and brings them to the attention of planners and decision makers
- projects can benefit from the experience and knowledge of a wide range of specialists from in and outside of the region
- modest support from appropriate external assistance groups, when added to national commitments, can serve as a catalyst in generating support.

Establishing regional projects and programs require that there be a credible regional body or organization to promote, implement and monitor the activities. Regional networks made up of participating institutions, and headquartered at one of these institutions are one of the simplest and most effective mechanisms for this.

SMALL-SCALE MINING WORLD-WIDE --TOWARDS NEW DIMENSIONS OF DEVELOPMENT

D N Bhargava
Indian Bureau of Mines, Nagpur, India

INTRODUCTION

It has been my privilege to have been associated with three of the four international events which took place in the past eight years, relating to small-scale mining. I participated in the first ever international conference held in Nov-Dec. 1978 at Queretaro, about 100 km from Mexico city. It was organised jointly by the United Nations Institute for Training and Research (UNITAR) and Secretaria de Patrimoniy Fomento Industrial (SEPAFIN) and Consejo de Recursos Minerals. I missed the second conference held in Africa 2-3 years later. But I attended the remaining two international events, namely the Workshop on Mineral Policy for Small-scale Mining held at Vigyan Bhawan, New Delhi, from 15th to 17th November 1984 at the Round Table on Small-scale Mining held during the XII World Mining Congress, New Delhi from 19th to 23rd November 1984. The Workshop was jointly sponsored by the Regional Mineral Resources Development Centre (RMRDC) of ESCAP and the Ministry of Steel & Mines, Department of Mines, Government of India.

The first conference provided valuable realisation about small-scale mining as a world-wide phenomenon, present almost in every country whether developed or developing or under-developed with close similarity in its characteristic features, problems, hopes and aspirations. The world itself appeared small and it gave a warm feeling of kinship among the 178 delegates drawn from 63 countries. Seventy two papers were presented. The week-long deliberations gave a good insight into the state of small-scale mining throughout the world, and the Conference concluded with recommendations for three sets of action addressed to the Governments of various countries of the world, to the traditional sources of bilateral and multilateral technical and financial assistance including the agencies of the United Nations and to the small-scale mining fraternity itself. The full text of these recommendations is placed at Annexure I. It may be said that these recommendations almost became the harbinger of new dimensions of development for small-scale mining everywhere.

The narration of experience already gained in some countries provided valuable ideas for adoption elsewhere. The Workshop and Round Table on Small-scale Mining (1984) served as occasions for stocktaking, further exchange of experience and for renewal of commitment to the cause of small-scale mining.

I propose to provide glimpses of some salient features of small-scale mining in different countries as presented at the Mexico Conference, the constraints experienced generally and the collective opinion of the international small-scale mining community about the new dimensions envisaged for development.

SMALL SCALE MINING - SOME INTERNATIONAL EXPERIENCES

In Bolivia, small-scale mining is being extensively undertaken for increasing the employment potential and for producing mineral commodities needed for local and indigenous needs. In their view, small-scale mining sometimes becomes a precursor of large-scale operations if it leads to the discovery of very large reserves. The Government of Bolivia has for this purpose a technical agency which provides facilities for the evaluation of small mineral deposits as well as for providing financial assistance to small entrepreneurs. This scheme has been successful in accelerating the development of small mines in Bolivia.

Canada also subscribes to the view that small mines often lead to the discovery of large deposits. A private marketing group in Canada has been providing common facilities of technical and financial assistance for developing small mines. It holds the view that developing strong marketing facilities for buying mineral concentrates from such groups engaged in small-scale operations and assuring their fair prices is essential; the buyer may sell the blended products to consumers of large quantities. SOQUEM, an agency of the Government of Quebec, Canada, serves as a vehicle for the promotion and active development of Quebec's mining industry. SOQUEM consists of three techical departments - Exploration, Miing, and Research. The planning and execution of joint ventures with private industry is carried out by management committees the members of which represent the parties sharing the cost of the projects. SSOQUEM has demonstrated that Government-owned Corporations can be as e fficient as those in the private sector. It has brought together the best brains, modern technology, available knowledge and ingenuity with success.

The Government of Chile has established Government-owned agencies for buying and marketing of minerals. It also promotes and provides assistance for setting up beneficiation plants and even smelters. It buys minerals in the raw form from the small mine operators ensuring them payment of fair prices, gets the minerals beneficiated or smelted. The Government also provides the infrastructure, facilities for mineral analysis, and loans for purchase of equipment.

In Finland, metalliferous ore deposits as small as 20,000 to 200,000 tonnes are mined and ore is transported to distant ore-dressing plants. If the product of the mineral reserve and the distance between the mine and ore-dressing plant is less than 10 million tonnes-km, setting up an ore-dressing plant at mine site is not considered worthwhile. For deposits in case of which the "product" exceeds 10 million tonnes-km, semiportable ore-dressing plants with capacities of about 200,000 to 300,000 tonnes per year are moved to different sites as desired. Some small mines are worked as satellite mines to the bigger mines in the neighbourhood. For exploration, even amateur prospectors are encouraged. The activities of such laymen have resulted in many important finds. Cash awards are given for the best samples.

In Japan, small pegmatite deposits usually 200 m. long, 1 to 2m thick and traced to 100 to 200 m down the dip are exploited on small-scale for recovery of felspars.

In Kenya, the Government's policy is to encourage small-scale mining by private prospectors and entrepreneurs and mining co-operatives. Small-scale mining supports a high percentage of unskilled cadres. Gemstones are the basis of small-scale mining in Kenya and tourists are the major buyers of the locally processed gemstones.

In Mexico, small-scale mining is considered as providing an avenue for more employment, for obtaining marketable minerals from small deposits at commercial prices for meeting Mexico'sown needs and sometimes for exports too. It is also viewed as an advance tool for exploration. There are two agencies in Mexico for encouraging small-scale mining: (i) Commission-de-Fomento Miners and (ii) Consejo-de-Recursos Minerals. The first agency provides technical and financial assistance to small-scale mining operators. The latter provides a strong base for the national exploration programme. It discovers the mineral deposits and develops them into small operating mines. The former provides the necessary incentives for these working mines to be taken over by private entrepreneurs or operate them through joint ventures. Fiscal and technical stimulation is continued by these agencies even after the mines become fully operational.

Even in the USA, 80% of the mines are categorised as small. The Government felt the need and provided incentives for search of minerals such as tungsten, chromite, manganese, mica and uranium and for the development of small mines. Under one such programme, financial assistance was provided for mineral exploration. It led to the discovery of minerals worth ¢ 1.3 billion. The Government costs were $ 41.0 million and of these $ 7.6 million (18.5%) were recovered as 5% charge on gross proceeds from ore sales of $ 152.4 million. The Government also arranged to purchase the minerals from small mine owners.

After Zaire became independent in 1960, its small-scale mining for gold, cassiterite and diamond got discontinued. With a view to revive it, a new Mining Law was promulgated in 1967. It permits the active participation of Zairian nationals and of the Government itself in improving the national mining resources. Direct exploration of minerals by skilled miners has been stipulated. The Government has set up agencies to buy diamonds from small-scale miners. Similarly, agencies for purchase of gold and cassiterite were also envisaged.

These are just some salient examples of small-scale mining in vogue in various countries and are illustrative of its being practised world-wide.

MINERAL POLICY FOR SMALL SCALE MINES

The Workshop on Mineral Policy for Small-scale Mining (1984), attended by delegates from Asian and African countries, brought out the similarity of experiences with small-scale mining in various countries as also the nature of problems encountered. The discussions provided excellent opportunity for the exchange of views among the participants. The salient observations and recommendations of the Workshop are as follows:

(i) Many mineral deposits do not lend themselves to large-scale mining and the increasing difficulties in mobilising large investments needed for the development of large mines due to uncertainties of long-term mineral markets also favour small-scale mining operations.

(ii) In almost all the countries represented at the Workshop, the small-scale mining sector suffered from lack of proper evaluation of the mineral reserves. It was felt that this inadequacy must be removed to enable the small mine owners to plan mining operations on a scientific basis. The consensus at the Workshop was that detailed exploration of small deposits must be done by some developmental agency of the government and the full report should be made available to the entrepreneurs free of cost or at concessional charges. This will encourage the growth of well-planned small mines. Joint mining enterprises could be launched in which the Government may retain control through majority share holding.

(iii) It was felt that small-scale mines in almost all the participating countries have a dearth of qualified mining engineers, geologists, supervisors and even properly trained workers. The consensus was that in the present situation, the Government should provide the expertise for the evaluation of mineral deposits, preparation of schemes of mining in a systematic manner, and for guiding the mining operations from time to time. The small mine owners should have the obligation to work according to the scheme and follow all the relevant statutory provisions.

(iv) It was observed that in almost all participating countries the mining lease areas of small mines were generally small and insufficient for proper organisation of mining operations and the period of lease was not long enough to encourage sufficient investment. The consensus was that the area granted under mining lease should not only cover the mineralised ground but enough area for development of mine workings, disposal of waste, locating functional and residential buildings, etc., and the period of lease should be long enough to attract adequate investment. Also care should be taken to see that the small mines should have enough size to be planned for such a scale of production which will provide sufficient turnover to enable the respective mines to arrange for the observance of all applicable statutory responsibilities. Preference should be given to qualified geologists and mining engineers in the grant of operating rights.

(v) Some agencies - Governmental, co-operative or private should arrange for lending such equipment which is essential for efficient mining but is required only intermittently but the small mines such as bulldozers, front-end loaders, blasthole drills, etc.

(vi) It was felt that it would be useful if each country prepares a catalogue of machinery manufactured by it for use in small mines. A compendium of such catalogues will be useful to all countries. The profession of consulting mining engineers/geologists, etc. should be encouraged to provide part-time expert advice from qualified personnel.

(vii) Government agencies at the national and regional levels should provide laboratory services for mineralogical studies, analysis of ores and minerals and beneficiation tests. A portion of the revenue from the mining industry should be kept aside for the benefit and development of the industry.

(viii) It was noted that in most countries, safety regulations were common for all mines, big or small, but in several countries provision did exist for exemption from a number of provisions for certain categories of mines. It was recommended that simplified versions of the safety regulations may be prepared for better understanding of safety requirements by managers, supervisors, and workers in small mines.

(ix) Government agencies should undertake market studies particularly of minerals mined by small producers and widely disseminate the results of such studies in a form easily understandable by small-scale miners. It should collect market intelligence and should disseminate this information for the benefit of small-scale miners through news letters or other appropriate means. It should also review the standards prescribed for minerals used by various consuming industries in order to make the standard more realistic taking into account the composition of the minerals available.

(x) Governments should identify or create agencies that can provide support for development of infrastructure for small-scale mining, provide fiscal incentives for support of the infrastructure development to achieve national goals and provide to clustres of mines, wherever they come into existence, necessary infrastructure for their development.

(xi) Emphasis was laid on the need for creating training facilities for miners employed in small mines to develop skills in various mining and allied crafts. Advisability of promoting the establishment of "grass roots" organisations was stressed, which would have good contact and credibility with small miners. These could be trade organisations, non-governmental bodies, co-operatives, and so on. Such organisations could provide effective route for disseminating knowledge and information from international and national sources to small miners.

The Round Table held at XII World Mining Congress was also largely attended and the main conclusions were as follows:

(i) Small mines sector can be a vital part of every country not only for its own production and contribution to the economy but also for its role in the creation and continued support of major mining.

(ii) The importance of small mining sector should be recognised by both the national government and the local community. The important role of the small miner - as an individual - should also be recognised.

(iii) Mining leases should be long and large in order to give real security and should only be terminated if the small miner fails to operate the deposit.

(iv) An organisation to help small-scale mines to finance the initial efforts in exploration and development and to provide technical guidance needs to be created.

(v) In the early stages, arrangements should be made to provide a guaranteed price for the small miners' output either as cash or as a direct offset against initial loans given to help development.

(vi) Co-operative services and technical assistance should be available to help the small miner to improve his efficiency and production.

(viii) Facilities for training small-scale miners need to be organised.

(viii) An organisation to establish beneficiation facilities in strategically located places needs to be created to upgrade products of small-scale mining operations.

It is interesting to note that the conclusions arrived at the Round Table of XII World Mining Congress more or less were the same as the conclusions and recommendations of the Workshop of Asian and African Countries. This indicated harmony in the world-wide view of small-scale mining.

SMALL SCALE MINING - THE INDIAN SCENARIO

Looking at the Indian scene, it is observed that over 95% of the mines in India are small when considered on the basis of the two main characteristics which differentiate small-scale from large-scale mining, namely:

(i) The absence of the regular use of deep blast hole drilling, heavy blasting, or heavy earth moving machinery.

(ii) The limitation of employment to 400 persons in all in the mine or exceeding 150 persons in workings below ground. (A number of mechanised mines may conform to this limitation of employment but they are not classified as small due to the mechanised method of mining adopted.)

In the light of this position, it would be interesting to consider how India stands with regard to the various observations and recommendations made at the Workshop on Mineral Policy (1984). This would indicate the developments that have already taken place as also the new dimensions of development towards which progress would be called for in future. India's position emerges as follows:

(i) Adoption of small-scale mining in India is due both to the fact that some mineral deposits lend themselves only to small-scale mining and also that it is possible to undertake them without large investments.

(ii) In India too, the small-scale mining sector has suffered from lack of proper evaluation of the mineral reserves and thus from lack of planning of mining operations on a scientific basis. Effort is however on since the last 15 years to remove this inadequacy. Technical consultancy is being provided by the Indian Bureau of Mines at highly concessional rates for geological mapping, preparation of schemes of further exploration and preparation of schemes of mining. But the demand has mainly remained for geological mapping and rough appraisal of mineral resources. Over 250 assignments have been completed. The IBM is expanding its Technical Consultancy Services for taking up the formulation of schemes of mining. The industry is welcome to take advantage of this service. State Directorates are also undertaking exploration and formation of joint sector enterprises has taken place in states like Andhra Pradesh, Bihar, Gujarat, Kerala, Maharashtra, Rajasthan and West Bengal. It is hoped that this development will catch on in other states too.

(iii) There is dearth of qualified technical personnel in small mines in India too. Although the position has improved to some extent due to the induction of diploma holders in mining, the need for association of experienced and qualified mining engineers is necessary for preparation of schemes of mining and occasional guidance. The newly amended Mines and Minerals (Regulation and Development) Act makes it compulsory that new entrepreneurs first carry out exploration and prepare schemes of mining before grant of mining lease could be considered. This should help in induction of mining consultants for such services and to improve the position of availability of qualified personnel to the small-scale mines too. Although there are vocational training centres for training of mine workers, they essentially deal with "do's and dont's" of safety. There is still wide scope for development of training centres for mine workers to teach crafts and skills. The IBM proposes to expand its Training Centre to provide courses for front-line supervisors. Training centres for workers will have to be planned to function in well known centres of mining activity. Provision may have to be made for on-the-job training of workers under the supervision of trained supervisors.

(iv) Mining lease areas are generally small and lease period is sometimes short. The exploration of the area by the entrepreneurs and preparation of the scheme of mining prior to the grant of mining lease would ensure that the area required will be duly determined basing it on the viability for mining; therefore it would no longer be adequate. It is expected that with a scheme of mining available to the authorities for consideration, the period of lease allowed would also be adequate. If the applicants are themselves qualified geologists or mining engineers, the question of the applicant not being able to afford the services of qualified persons would not arise, and it is hoped that the mining operations under the own supervision of such personnel would be more systematic and in accordance with law.

(v) Some State Government agencies have schemes for providing mining equipment on hire but this practice has not gained popularity. It calls for a view to discover the reasons for the limited demand. It would be desirable for the industry to make it attractive to private agencies to offer such services so that they do not have to look to the Government for support in this regard.

(vi) In India, almost all the machinery required for small-scale mining is available indigenously. The IBM publishes periodically a hand-book of such machinery. The hand-book is very popular among the mining industry. It is in demand in some other developing countries too who wish to have a knowledge of the range of mining machinery being manufactured in India.

(vii) The IBM has three laboratories at Nagpur, Ajmer and Bangalore which undertake mineralogical examination, chemical analysis and beneficiation tests. These services are offered at attractive rates. The facilities are being utilised both by private and public sectors, including the small mines' sector. Besides, some state directorates of geology and mining, some CSIR laboratories and some private laboratories are engaged in one or more of these services. The IBM would be willing to expand its capacity by opening new laboratories if adequate work-load is identified to justify their setting up.

(viii) For the present, there are common safety regulations for all mines but a number of provisions are not applicable to small mines under certain conditions. Even so, the question of making separate legislation for small mines has attracted the attention of the Government.

(ix) The IBM is undertaking market survey studies for minerals including those which are mined wholly or partly on small-scale. The minerals covered by these studies include Barytes, Bentonite, Chromite, Granite, Kyanite & sillimanite, Manganese dioxide, Rock phosphate, and Talc/Steatite. The studies are published as bulletins under "Market Survey Series" and are available for free sale at nominal cost. IBM proposes to s trengthen this activity.

(x) Identification or creation of agencies that can provide support for development of infrastructure for small-scale mines, for providing fiscal incentives, etc. are the new dimensions for development to be pursued in future. India does have a number of incentives offered to small-scale industries but unfortunately these are not extended to small-scale industry, The Federation of Indian Mineral Industries has pleaded strongly for small-scale mining to be considered at par with small-scale industries. This demand is being carefully examined. It is hoped that some practical ways would be found to provide support for infrastructure development as well as other financial incentives.

In conclusion, it may be said that India has a sizable small-scale mining industry. It is receiving support in many ways from Government agencies by way of technical services at concessional charges. The deliberations at the international forums have served to identify the new dimensions of development necessary for the health and viable growth of small-scale mining. The recommendations of these conferences have received sympathetic attention in our country. It is hoped that through the support being provided kby the Government and its agencies and through measures of self-help by the small-scale mining sector itself, the industry could look forward to a bright future.

ANNEXURE - I

RECOMMENDATIONS OF INTERNATIONAL CONFERENCE ON THE FUTURE OF SMALL-SCALE MINING QUERETARD,MEXICO 27TH NOV to £TH DEC.1978

A. To Government entities the following program is recommended to assure maximum utilisation of small deposits:

1. Establishment of funding mechanisms designed specifically to assist in the exploration for mineral deposits and the development of small mines.

2. Provision of direct technical assistance to small-scale mining as follows:

   (a) Geoscientific and geotechnical information and related assistance.

   (b) Advice on proper mining and processing methods.

   (c) Laboratory and metallurgical analyses and assay services.

   (d) Equipment pools or other means for making the most suitable equipment available, and

   (E) Custom concentration plants and/or portable mills.

3. Additional required financial support in such forms are:
   (a) Establishment of special development funds for small-scale mining.
   (b) Tax incentives to encourage development of small deposits, and
   (c) Assistance in marketing.

4. Provision of necessary infrastructure, where justified, such that the small-scale mining industry receives the benefits of access roads, water and power supply, medical services, and schools.
5. Provision of special training facilities for the small-scale miner.
6. Promotion of safe and hygenic working conditions, and
7. Education of the small-scale miner on preservation of the environment.

B. The Conference calls upon traditional sources of bilateral and multilateral technical and financial assistance, keeping in mind the important role technical co-operation among developing countries (TCDC) can play here, to support Government requests for assistance to their small-scale mining sectors through:

1. Strengthening of Government services relating to:
(a) Legislation, taxation, financial and other promotion.
(b) Exploration, research, development and mining, and
(c) Beneficiation, smelting, processing and marketing of metallic and non-metallic mineral resources.
2. Advice and assistance for the constitution and operation of:
(a) National exploration funds and landing institutions.
(b) Small mining enterprises and cooperatives; and
(c) Specialised training schools.

3. Encouragement for the sub-regional, regional or global exchange of information on methodologies and technologies applicable to small-scale mining and mineral processing through the organisation of workshops and training seminars.

4. Assistance in exploration and development, including the expansion of the terms of reference of the United Nations revolving fund for National Resources Exploration so that the fund can assist in the exploration also for small deposits, and the granting of loans by the World Bank for small-scale mining.

C. 1. The Conference proposes the continuance of the program of world conference on Small-Scale Mining on a biennial basis.

2. The Conference suggests the establishment at an early date of an International Research Institute on Small-scale Mining dealing with all aspects of research and development of small deposits and small-scale mining.

***

STRATEGIES AND TACTICS FOR STIMULATING MINERAL EXPLORATION AND DEVELOPMENT IN DEVELOPING COUNTRIES

Tej Bhan Malhotra
Consulting Mining Engineer,
New Delhi, India.

ABSTRACT

Mineral exploration includes regional topographical, geological, geophysical and geochemical surveys to indentify areas of mineral potential. Potential ore-bodies are then investigated in detail by drilling, trenching, pitting, exploratory mining and collection and testing of samples and a feasibility study is prepared.

Availability of geological information and expertise indifferent developing countries varies considerably. It is advisable that mineral exploration is organised by a country itself, if necessary through multi-lateral or bi-lateral assistance, avoiding use of multinational companies for this work.

Certain basic objectives must be kept in view in mineral development. These are safety and health of personnel, no avoidable wastage of resources, protection of environment and optimum benefit for the country from its resources.

Where expertise and capital is available, mineral development should be done under local control. Otherwise foreign collaboration may be obtained. But such a partnership will endure only if it is mutually beneficial and is based on an unambiguous agreement, that is implemented in good faith.

Important factors for stimulating mineral development are the existence of good resources, expertise, capital, infrastrcture, market, an attractive fiscal regime and good law and order and working conditions.

INTRODUCTION

As minerals are an important and irreplaceable part of a country's heritage, the goal of mineral development should be to obtain optimum benefit for the country from the present and future use of its minerals.

Mineral development consists of a series of activities, which include exploration and discovery of minerals, their appraisal and evaluation, development of mines, production, trading and utilization of minerals, ores and metals and reclamation and rehabilitation of land from which minerals have been extracted.

MINERAL EXPLORATION

The first activity in the mineral development system is that of exploration. Mineral Exploration is a complex activity, which begins with broad geological surveys, consisting of regional topographical, geological, geophysical, geochemical and remote-sensing surveys. These are designed primarily to identify areas of mineral potential. Areas thus identified then need detailed exploration, which includes pitting, trenching, drilling and collection and testing of samples. A pre-feasibility study is prepared at this stage to determine the technical and economic viability of the deposit to decide whether further detailed investigation of the deposit is justified.

Potential ore-bodies are then investigated in detail to determine the technical and economic parameters under which the deposit could be developed. This investigation may require more drilling, trenching, pitting and underground exploratory. mining, as well as collection of bulk representative samples and their metallurgical testing in laboratories and pilot plants. A technical and economic Feasibility Report is then prepared, outlining definite working plans and capital and operational cost estimates, on the basis of which an investment decision is made.

It will be seen from the above, that mineral exploration is a highly complex, high-tech operation, which is expensive and time-consuming, and involves lot of field work. Other things being equal, a competently carried out exploration work can make the difference between a valuable discovery and anunsuccessful programme.

The basic geological information as well as expertise needed for mineral exploration that is available in different developing countries varies considerably. In some countries, this is quite large, and only limited assistance may be needed in some areas of some new technologies, whereas in some other developing countries, this may be very meagre or none at all. The entire work of exploration may thus have to start from the beginning. Further, some developing countries may have no surplus resources available for carrying out regional geological studies and exploration.

As investment made in exploration may or may not find economic mineral deposits, it is a high-risk venture. If private enterprenours, specially expatriate companies, have to take part in such ventures, they would be generally looking for extra-high benefits, incase they succeed in finding an economic deposit, as they would like to compensate themselves for the infructuous expenditure incurred by them elsewhere, where they fail to find an economic deposit. The exception to this general rule may be, where the private company involved is more interested inassuring itself a regular supply of certain important raw materials, or in maintaining its monopoly position in some mineral. The out-go of extra-high financial returns in natural resources field is not likely to be popular in any country today, and should therefore be avoided.

The strategy that a country may adopt for stimulating mineral exploration will depend upon its circumstances and its economic and political philosophy. But as is indicated above, the best solution as far as organisation of basic geological studies and mapping and detailed mineral exploration is concerned would be for a country to organise this work itself.

Where a country has the resources and the expertise to carry out prospecting and exploration work itself, the work could be divided in two parts - that of basic geological studies and mapping, which should be carried out by the national or state geological surveys and allied organisations, and detailed exploration work, which may be carried out by the mining organisations. The latter work could also be carried out by specialised organisations on contract for the mining organisations.

Where the developing country lacks resources and or expertise in the exploration field, it may consider using the following alternatives for the purpose.

(a) Technical co-operation amongst developing countries,

(b) Bi-lateral aid,

(c) Financing through bi-lateral aid, and expertise through most suitable sources,

(d) UNDP/TCDC technical assistance,

(e) United Nations Revolving Fund for Natural Resources Exploration.

(a) Technical co-operation amongst developing countries (TCDC) holds great promise for the future, and may be of great permanent benefit to the countries concerned. It will be more attuned to success and less costly in less sophisticated situations. Many a time, assistance is really needed, not of highly qualified professionals, but of sub-professionals and technicians. These could be more usefully provided by one or other of the developing countries. Again, personnel from developing countries are likely to adjust to work in other developing countries more easily. It is recommended that U.N.Regional Commissions should take special interest in organising co-operation amongst developing countries in this field.

(b) Bi-lateral aid, if given sincerely, can be a very useful source of assistance in this field. In one country, where this author had the privilege to work with United Nations, a programme of geological surveys, mapping and mineral exploration has been going on since several years through bi-lateral aid. It is one of the most successful programmes in this field that was ever launched. Excellent geological work has been done, and a large number of valuable mineral deposits have been discovered, which are already making valuable contribution to the economy of the country.

(c) Financing through bi-lateral aid, and expertise through most suitable sources: While some donors insist on sending their own experts for working on the projects being funded by them, there are instances, where the donors have adopted a more liberal attitude and allow the recepient country to have the expertise from any place of its choice. This can be very useful indeed for the recepient country. It will certainly be a good example to follow by various aid-giving countries and consortia.

(d) UNDP/TCDC Technical Assistance: This has been a very useful source of assistance in this field for the last several years in many countries. Some very useful bi-lateral aid programmes has also been co-ordinated with some such U.N. assisted programmes.

(e) United Nations Revolving Fund for Natural Resources Exploration: This is another useful programme being administered by United Nations for the last several years, and has been doing very valuable work.

While various aid programmes should make their contribution to the economic uplift of the country, the recepient country's decision-makers should not forget their responsibility in selecting and accepting the programmes and the experts. Generally such programmes and experts are expensive, and the recepient country should make every endeavour to ensure that they are getting the best benefit from the aid that they are receiving.

Whatever the stage of a developing country's level of geological information, the following tactics may be found to be useful in discovery of new mineral deposits:

(i) Attractive awards may be given for bringing in informationor samples of minerals, old slag heaps and old workings, may lead to discovery of economic deposits. This suggestion may appear to be elementary and not sophisticated enough to warrant attention. It may be noted however, that except for a few mineral deposits, nearly all the presently known mineral deposits have been discovered in India from old mine workings. It is understood that the best lead-zinc deposit known in India today, was discovered only a few years back, not by aerial geophysics that had been carried out earlier in the area with foreign expertise, nor by the geological studies carried out by the national or state geological survey organisations, nor even by the nearby located public-sector organisation, which is the sole organisation responsible for exploring, mining and processing lead and zinc in the country, but through examination of old slag dumps in the area, which had come to notice in a completely different context. It is agreed that a systematic approach is necessary and important in search for minerals, but one cannot ignore the fact that a number of mineral deposits being worked in the world today were found by accident, and as systematic geological work cannot be completed in a hurry, there will be many opportunities for discovery of new mineral deposits in many places till the completion of this work.

(ii) The discoverer, or if he is not interested, the first applicant for mineral concession for a deposit, not held already under a mineral title, or not under current investigation by a Government organisation, should normally be granted such a mineral concession. And he should be able to hold it as a right to property, which he can sell to approved parties. Many objections can be voiced against this suggestion. It may be argued that an applicant may have no expertise or financial resources todevelop such a deposit, and it will not be in national interest to let a mineral deposit lie undeveloped. The Government may provide minimum standards of technical expertise and financial resources required for developing such deposits, as also steeply escalating minimum royalty rates, to make it uneconomic for anybody to hold mineral properties undeveloped for long. Other difficulties can similarly be overcome, if tackled in good faith. Mineral prospecting has its own fascination for some people who will not be happy doing other jobs, even if they are financially more rewarding. It is suggested that this healthy fascination may be used for the benefit of the country and such people.

If the country's policies do not permit the grant of mineral concessions to private parties, he should be handsomely rewarded otherwise. Again some local people may have information about some mineral occurrences, which when scientifically explored, may turn out to be good deposits. Every effort should be made to derive benefits from such knowledge.

MINERAL DEVELOPMENT

The development of mineral resources should be integrated into the total development plans of the country, and not considered as an isolated economic activity. While substantial benefits can accrue to a country from a properly organised and managed mineral industry, it should be remembered that a poorly organised mining industry can also have adverse effects on the economy of a developing country.

To achieve the goal of mineral development, as enunciated in the beginning of this paper, certain basic principles must always be kept in view. These include ensuring safety and health of those engaged in the industry, no avoidable wastage of resources in their extraction and use, protection of environment, assuring adequate supply of minerals for country's needs, balanced development and diversification of national economy through increased mineral processing and mineral-based manufacturing, obtaining best returns from exports, and production and utilization of minerals with greatest possible efficiency. This is being emphasised here at length as it is necessary to keep these basic principles in view always, when considering the strategies and tactics for stimulating mineral development.

Important facts for stimulating mineral development in a country are the existence of good mineral resources, and availability of expertise, financial resources, infra-structure, markets, an attractive fiscal regime, and of course good law and order and working conditions.

The Government has to take a policy decision whether mineral development will take place in private sector, public sector, or in a mixed economy, and whether foreign capital will be welcome for this work in any form.

Where the country has the resources and expertise for mineral development, it is better to avoid the use of foreign capital for the purpose. The memories of foreign ownership of natural resources in various countries are generally not that happy. The record of foreign mining companies in responding to the needs of the developing countries has been generally dismal. In fact, they rarely showed even an awareness of the needs of the developing countries.

When the national policy is to allow both public and private sector in mining industry, equal facilities should be given to both sectors. The strategy should be to improve efficiency of the industry through competition. It is clearly understood that the basic principles of mineral development enunciated earlier will be strictly followed by both the sectors.

If the state policy is that mining has to be in public sector, it is better to have more than one public sector corporation in each major mineral or group of minerals, to encourage competition for improving efficiency. Where feasible, mining should be allowed both in public and private sectors. By the very nature of their requirements like the need to ensure public accountability in their operations, and high overheads, it may not be economical for public sector units to operate small mining enterprises. It will stimulate mining development, as well as development of remote and economically backward areas, if small mineral deposits are allowed to be developed by private enterprise. For large projects, specially those requiring lot of foreign exchange, a country could obtain loans from multilateral sources like the World Bank institutions or bi-lateral sources, say for purchase of equipment, or sometimes even from commercial banks.

When a country lacks capital and expertise for its mineral development, it may have to depend upon foreign sources for the same. As in any market situation,. one will have to look for a source of capital and expertise, in whose interest it may be to enter into such a partnership. Foreign enterprises may be interested in such a partnership for assured and long-term supply of certain minerals, or to improve their position of control on World supply of some minerals, or only as a business proposition, or sometimes mainly to secure a presence in the country for possible larg er benefits later.

The country must keep in view the basic principles of mineral development, while negotiating with a foreign enterprise. It is most likely that the objectives of the Government and the foreign enterprise will diverge considerably. The foreign enterprise will normally be expecting very high financial returns, much higher than what they would be content within their own or other developed countries, to offset the risks that they visualise in the developing countries.

The objectives of a foreign enterprise may be stated as under:

(a) Safety and security of personnel through proper law and order.

(b) Reasonably comfortable living conditions for their personnel.

(c) Safety of capital.

(d) Good financial returns, consistent with risks involved and opportunities available elsewhere.

(e) Availability of minerals at reasonable prices on a long-term basis.

(Sometimes a foreign enterprise may be mainly interested in this only, and may be interested in financing the project through loans against assured future supply of minerals, and may not be interested in actual management of the project).

In addition to the basic principles enunciated earlier, the objectives of the Government may include the following also:

(i) Earning or saving foreign exchange

(ii) Creating employment, directly or indirectly, often in remote areas.

(iii) Promoting backward and forward linkages to maximise value added within the country.

(iv) Regional development.

It is suggested that the two parties should strive for a partnership that is mutually beneficial and can endure.

While safeguarding fully their basic interest and essential requirements, the developing country should reduce other risks visualised by the foreign enterprise to a minimum, establishing for themselves a reputation for fair and clean dealings. This will encourage competition amongst potential investors, resulting in fair and better terms for the country.

Good faith and strict adherence to an agreement that has been freely negotiated, with full knowledge of facts with both parties, are the basic ingredients for the success of such an agreement, and a successful enterprise of this type will be a good stimulant for further development of other areas in the country.

The agreement should be clear and unambiguous. The state should have share in equity of the enterprise (in return for grant of mineral rights, without having to pay for it in cash), as well as representation in the Board of Directors for an effective say in policy matters, as also representation in a Management Committee to be well-informed about the operations of the enterprise, but at the same time, there should be no unnecessary interference in day-to-day management affairs.

The share of Government in equity and taxes will be negotiated and will depend upon the value of the deposit. To avoid controversies later, the agreement may provide that in case the profitability of the enterprise proves to be better than that estimated in the feasibility report, the Government will get a higher share of the higher profit.

As mining is a basic industry, the Government must be ever alive to its needs and problems. The interests of an enterpreneur (individual or a local or foreign enterprise) should ideally coincide with the national interest. Even a take-over of some mineral areas by the State may be justified and necessary in national interest sometime, but it must be understood that expropriation without fair and reasonable compensation can certainly not be in national interest. An enterpreneur should not cavil, if for better and optimum development of a mineral area or a deposit, it is taken over by the Government, say for example, a mineral deposit which was originally thought to be a small deposit, but has later turned into a large deposit requiring large capital and other facilities. But it is essential that a fair and reasonable compensation must be paid for it, with perhaps a bonus as a gesture for enterpreneur's services, which have turned it into a large deposit. To do otherwise will be a disincentive for proper and scientific mineral development. The developing country should ensure that its nationals are being trained at all levels to be able to man the industry efficiently in course of time. As the local personnel gets trained, the expatriate staff will gradually retire. This replacement should be done in a fair and graceful manner, with due regard to the needs of the enterprise.

As exploration and mining technology, as also instruments and equipment are being constantly improved, the mining industry must keep itself abreast of progress, through its R&D activities, literature surveys, visits to advanced mining facilities elsewhere and through bi-lateral and multilateral cooperation. Any improvement in efficiency will reduce costs, and stimulate further mineral needs and development. Research and development efforts in finding new uses of minerals will have a similar effect.

HUMAN RESOURCES DEVELOPMENT FOR THE MINING INDUSTRY IN DEVELOPING COUNTRIES - CHALLENGES OF THE 21st CENTURY

BY

Professor R.D.Singh
Kothagudem School of Mines
Osmania University
Kothagudem, India.

ABSTRACT

Mineral resource of the third world countries make an impressive inventory but the human resources, though large, are not fully developed to match the tasks set by the mineral industry to cope with the fast technological and social/political changes. The paper briefly describes the set up thus far generally adopted to develop this manpower for the mining industry and draws out the weaknesses of the system. Drawing upon past experiences and forecasting the future needs, the paper defines the steps that may be taken to develop human resources for the mining industry to meet its varied requirements. Finally, the paper emphasises that to assure the growth and prosperity of the mineral industry, the LDCs must quicken the pace of development of human resources so that the technological gap between developed and developing countries is narrowed or eliminated in the shortest possible time.

INTRODUCTION

The non-fuel mineral resources of developing countries are impressive. The programmes for mineral development in the LDCs are equally ambitious. A World Bank study entitled "The Outlook for Primary Commodities 1984 to 1985" (Mining Journal, 1985) has underlined the increasing role played by the dev eloping countries in the consumption of non-fuel minerals. These shifts have been most rapid and evident in the case of metals and minerals, fertilizers and agricultural raw material. The World Bank report highlights the example of developing countries' share of World Consumption of metals/minerals and fertilizers which increased very rapidly between the mid-1960s and the end of 1970s, in some cases more than doubling. Export of fertilizers from the LDCs are forecast to show the most rapid change of all commodities considered-up from10% of the world total in 1970-81 to 26.4% by 1995. Zinc trading patterns will reflect the trend towards local processing of concentrates in the LDCs and the rising dependence of some industrial countries on metal imports from other countries (developing and industrialised).

More than 80% of the projected increase in world fertilizer production over this forecast period is likely to take place in the developing countries and Centrally Planned Economies. Growth will be most rapid in the LDCs where the rate is expected to average 4.7% over the period. By 1995, the developing countries are expected to account for some 43% of the World fertilizer consumption.

Evidently, the industrialized countries have declined in importance as centres of production of metals such as copper, steel, zinc, nickel and aluminium, whilst the regional shares of lead and manganese have remained rather more stable. Tin production, on the other hand, remains more concentrated in the developing countries. Perhaps the most dramatic shift has occurred in the case of chemical fertilizers where industrialised countries accounted for some 60% of the world output in 1964-66 but only 42% in 1982. The proportion held by the Centrally Planned Economies grew from 21% to 32% over the same period, while the production in the LDCs soared from 10% of the world total to 26%.

World zinc production is predicted to grow at an annual rate of 2.3% over this period and the degree of integration in the developing countries from mining to smelting should increase from 62% in 1974-81 to about 77% in 1995.

Side by side, considerable efforts have been put in by different LDCs to locate and work fossil fuels. Particular attention has been focussed to explore and exploit coal deposits so as toreduce dependence on oil. Mining methods in LDCs are generally labour intensive though the trend in recent years has been towards increasing degree of mechanization of the industry. Similarly, there has been a parallel trend towards increasing the number of openpit mining operations as those also offer better efficiencies than most of the underground ventures.

Newer techniques with sophisticated equipment and control systems are also finding application in certain countries. Automation and computer control are also envisaged. Fullest use of the potential of computers are projected in the mining industry.

Most of the LDCs have put to use the technologies developed in industrialised countries and in majority of the cases the technologies were introduced by colonia powers and managed by them. Even after the independence of different countries in the LDCs group, these countries depend on foreign know-how, foreign collaboration for the development of mining industry in their countries and in some cases the expatrates still occupy senior executive and technical positions and play major role in policy decisions as is the case in Zambia (Mining Journal, 1984). In some countries, e.g. Indonesia, as part of foreign government aid programmes foreign experts carry out geological exploration in that country. It will thus be abundantly clear that many LDCs depend on industrialised nations for expertise in their respective mineral industry development programmes. The system of education and training of manpower for the mineral industry has also been copied from the colonial powers. Many of the LDCs send their students abroad for higher technical education who return to their respective countries trained in a system which may not be relevant to their countries and the dependence on foreign expertise continues.

In the LDCs, the mineral industry scene is changing fast. The rapid growth of mineral industry calls for opening out of large mines perhaps with leaner ores which necessitates newer technologies for increased production at competitive prices. Also the situations created by the adoption of wrong technologies in the past need to be corrected and ore/coal retrieved as much as possible. This will call for technical personnel of the highest calibre.

Most of the LDCs have regained their independence in the recent past. This has brought in a change in the attitudes of labour who have become more assertive of their rights and have little deference to authority. The management must recognise this changes and prepare itself to cope with it. In other words, for the success of mineral industry appropriate management training becomes a pressing need.

To cope with the varied and fast changing requirements of the mining industry, human resources must be adequately developed not only to satisfy the immediate needs but also to enter the 21st century with confidence. The industry will need good managers to manage its business; it will need capable persons to make policy decisions and it will need men with imagination who can be innovators. Besides, the industry will need capable technicians and work force who will appreciate the intricasies and benefits of newer technologies, adopt them and assimilate them. The success of mining enterprises will demand trained and enlightened labour, technicians, managers, innovators, scientists, researchers and above all policy makers, who must be able to foresee and recognise changes and "see the change as an opportunity for improvement" (Deakin,1983). Information and education are essential prerequisite for the management of change.

PRINCIPLES OF MANPOWER PLANNING

The cardinal principle in manpower planning is to ensure that right type of men in the required number are always available to meet the needs of the industry. This would mean taking stock of the current position, foreseeing the future technological and expansion programmes and taking advance action to develop the men and the organisation and/or to recruit men from outside and develop them. Table 1 illustrates the processes involved in manpower planning emphasing the principle of integrated business and manpower planning. This will imply a capability to identify the future manpower needs in a qualitative and quantitative sense, and for educational and training centres to be structured and operated so as to provide for these needs. But forecasting of manpower needs accurately even in the short run is not possible. It should be emphasized that it is important to recognise the dynamic nature of employment, education and training and to be aware of the magnitude of the error inherent in forecasting. The qualitative aspect of manpower planning must also recognise adaptability of personnel and flexibility of employment, enabling manpower needs to be met inspite of what initially appears to be an adverse situation (UNESCO,1976).

Blanshard (1979) has opined that the "Category Job Analysis" is a valuable system of manpower planning and control and advocates that such a system is a must for every mine. Category job analysis can ensure that skills are properly employed. It can also indicate the need for recruitment and training. The system must be updated to cater to the requirement of changes. The underlying principle is to know early the manpower required in the various skills and to make sure that the best use is made of the available skills and to ensure advance action to train/retrain/recruit manpower for future requirements.

TYPES OF MANPOWER NEEDED BY THE MINING INDUSTRY.

Mining industry in every country, whether by statutory provisions or due to necessity, requires certain number of managerial and supervisory staff to be appointed at each mine. They are Managers, Assistant Managers, Engineers, Surveyors, etc. in addition to specialists. Besides, every mine employs a large labour force, clerical and administrative staff. Broadly speaking, the personnel in the mining industry may be classified into

TABLE-1

PRINCIPLE OF INTEGRATED MANPOWER PLANNING

	TODAY'S BUSINESS	TOMORROW'S BUSINESS
TODAY'S MAN-POWER	INVENTORY OF SKILL TO IDENTIFY	- Potential skill development.
	- Strength in skills	- Obsolescence of skills.
	- Absent skills, e.g. staff services, etc.	- Floating manpower.
	- Development needs Vs lateral recruitment.	- Which manpower to share.
	- Demographic mix to decide recruitment policies, etc.	
TOMORROW'S MANPOWER	New skill required.	- In company development of newly recruited manpower.
	Additional skills required at various levels	- Identifying fields of recruitment.

the following classes:

(i) The line and staff personnel including specialists,
(ii) Frontline supervisors,
(iii) Labour, and
(iv) Administrative and Clerical staff.

Managerial Staff

The source of the recruitment of line management for the mining industry is the educational system. The education and training of future mining engineers must prepare them to understand and act wisely in the changing industrial, social and political circumstances. Mining engineers of future will have to organise and lead a force that is not only more articulate but is also less concerned about work standards and probably less deferential to authority. Attitude towards work is changing; the view that work is worship is breaking down. Automation should provide a way of meeting the change and thus may become the major challenge of the professional engineer in the next century utilizing developments inmicroprocessing and other engineering techniques.

The environmental demands specific to mining will also call for an increased knowledge and expertise in, for example, the law relating to planning, the handling of people and the methods involving public enquiries and so on.

The technology available to both management and men is kchanging and is likely to continue to change. The engineers who are going to develop, appraise and use these new technologies must be professionally trained accordingly. It is difficult to prescribe the course content for the future requirements of mining graduates and specialists but there can be little doubt that study, to the appropriate level, inthe fundamental principles of engineering and science is the best way of ensuring that the industry will continue to have engineers who are able to exploit any future technology and technological changes to its full. The curricula should include a core of basic knowledge that will not become obsolescent together with training designed to develop adaptability, creativity, ability to evaluate alternatives and a sense of social responsibility. Te complementary curricula of a more applied nature will be more closely related to current needs, but it should also take into account the possibility of contiuing education programmes to meet the evolving needs of the technician or engineer in his employment.

Specialists

In these days of mining operations of increasing complexities, the management has to depend more and more on the mining specialists. In fact, the essense of modern management lies largely in the effective coordination of the work of the specialist team who will assist the management in solving the current and longer term problems (Moxham, 1979).

To perform his function satisfactorily, the specialist should be well trained in all basic aspects of mining, preferably with production experience and be fully conversant with all the statutory requirements applicable to the sphere of his work. But with the rising rate of technical and scientific innovation, the role of the specialist will change drastically in line with the scientific discoveries and the changing state of the mining industry. There is no doubt that the specialist will need training even in narrower fields than at present.

Frontline Supervisory Staff

Apart from training managerial staff and engineers, considerable attention must be paid to the training of frontline supervisors. They are the link between the management and the men. The result of the best project, of the best regulations, and of the best initiatives of management may come to naught if the frontline supervisors in their daily contact with the men lack tact and professional knowledge. These underofficials are a vital part of management organisation and have it within their power either to ensure success or jeopardise the success of any scheme which is set afoot.

These supervisors should be men of high character, just in their dealings with their subordinates and honest with their supervisors. A good supervisor should not be characterised by a hard and angry face but by a firm hand and a kind face. In addition, he must have a thorough professional knowledge. They should have a spirit of comradely helpfulness, inward discipline, soberness and complete trustworthiness. These kind of people do not grow on every bush but they have to be picked up with care and specially trained for their difficult job.

The techical school/polytechnics should design their programmes of instruction and training to turn out such supervisors. These schools should turn out the new entrants thoroughly built on theoretical and practical side of the job they are required to do. The schools should endeavour to shape the character of their pupils who should be fully trained to understand human values.

TRAINING OF WORK FORCE

It is increasingly being recognised now that the work force in the mining idustry must be adequately trained/retrained to be useful for the mining industry. Accordingly, suitable training programmes should be designed relevant to a particular situation in a particular country. For example, after the nationalization the collieries in the U.K. underwent massive reconstruction and mechanisation and the National Coal Board of U.K. provided opportunities for many thousands of its staff to undertake training courses either within or without its own facilities. Men in post and of varying ages were called to attend special courses run by the NCB for training in certain technical subjects. In the Lorraine coalfield, France has an excellent training institute at Forbach (Ecole Practique de Mine de Forbach) for training of mining personnel. In order to earn promotion, a worker must undergo training at this centre at different levels. In their report, the Technical Mission of the National Coal Board to USSR (1958) wrote that the training of junior officials, craftsmen and skilled workers generally was given great attention with the Soviet Mining Industry. In addition to preliminary training for new entrants, there were part time courses for a large number of basic underground trades. In India, Vocational Training Rules 1966 lay down the training schemes for certain category of staff. InZambia, the mining industry runs Training Institutes for imparting training to technicians. In the training scheme more emphasis is given on training on-the-job (Kumar, 1986).

The training schemes for the technicians and labour will have to be worked carefully keeping in view the technology adopted, the mining conditions and the social and political conditions of the people from where they come.

CONTINUING EDUCATION

Continuing education has always been an essential part of the career of any successful practising engineer or technician. The structure of formal continuing education is important, for its forms can range from a training seminar of short duration to a full-time course of many months duration. Different approaches are used in different countries. For instance, in the U.S.S.R. two or three months full time courses are usual, whereas in Western Europe short term courses of up to one week duration are more prevalent. A wide variety of institutional bases for continuing education of engineers and technicians have arisen. In some countries, the educational institutions have taken the lead while in some other countries individual industries or groups of industries may establish programmes, those in France being well known examples. Professional societies and individual unions have sometimes developed courses to meet the needs of their members (UNESCO,1976). Every country will benefit by providing continuing educational programmes for the mining industry which must be designed suitably to meet its own particular situation and needs.

CONCLUDING REMARKS

The mining industry in the LDCs is undergoing rapid technological and social changes. For its health and efficiency, therefore, human resources development must be effectively organised. As in many advanced countries, training of all categories of personnel (managerial, technical and labour) should receive attention. The training schemes should be dynamic and should be continually modified and updated and also to be effective the instructors should be updated at regular intervals. It will be desirable to arrange visits of instructors and selected category of technicians to advanced countries to study the modern developments intechnologies and the training scheme. The human development programmes may be designed keeping in view the developments that have taken place abroad already. But the education and trining of a particular LDC must be conditioned to be relevant to its own needs and must not be a carbon copy of what the developed countries have evolved, as to ape blindly will be counter-productive. Human resources must be aggressively and appropriately developed as on these resources the viability of the mining industry will largely depend.

REFERENCES

1. Kumar, S. (1986). Private Communications.
2. (1985). Mining Journal.
3. (1984)."Zambia 20 years on" Mining Journal, Vol.203, No.7784, pp. 285-288.
4. Deakin,A.F. (1983). "Management of change". The Mining Engineer, Vol.142, No.259, pp.555-556.

5. Blanshard, R.(1979)."Human Engineering and Personnel Problems in the Mining Industry". The Mining Engineer, Vol.138, No.210, pp. 651-55.
6. Moxham, J.A.(1979). "The Role of the Mining Specialist", The Mining Engineer, Vol.138, No.209, pp.587-93.
6. UNESCO (1976). Annotated Agenda for International Conference on Education and Training of Engineers and Technician.
8. (1958). Report of the Technical Mission of the National Coal Board on the Coal Industry of the U.S.S.R.,Part 2 - Training, Appendix 5, 56 pages.

OVERVIEW &
ROLE OF GOVERNMENT

MINERAL DEVELOPMENT IN LDCs - CONSTRAINTS AND CHALLENGES

A.K. Ghose
Department of Mining Engineering
Indian School of Mines
Dhanbad, India.

INTRODUCTION

Even if mining is a relatively small part of the total economic activity on a global scale, the ramifications of the mineral industry in a vast majority of developing countries have significant implications on national economies (Gluschke et al, 1980). The continuing recession in the world economy since 1980 and its attendant stresses on national balance of payment deficits, crippling debt-servicing problems, high unemployment and languishing raw materials demand pose serious structural problems of adjustment not merely for investments in exploration and exploitation in LDCs, they even threaten the political stability on a global scale. The state of African mining for instance, and of the sub-Saharan Africa in particular, is considered by many to be ominous in portent. The hydra-head of protectionism and trade barriers, as a part of an expedient resource diplomacy, also spell economic ruin for many mineral exporting LDCs.

In formulating any mineral development strategy for LDCs, one must not lose sight of the caveats of over-generalisation and over-simplification. Often loosely described as technologically dependent economies, the diversity of industrial structure and especially of the mineral industries in LDCs defy any universal prescription. Beneath the superficial similarity of the label of LDC, they differ vastly in size, shapes political hues and industrial base and there is nothing like a developing world perspective. In trying to develop a road map for the future of mineral development, we are not merely concerned with a road map, but road maps of each nation, each road as bumpy and as tortuous as situations permit. The basic purpose therefore should be to be devise policies that create desirable goals and to develop a persuasive thesis for mineral development in LDCs.

The paper will therefore seek to present a general analysis of the state of mineral economy in LDCs, highlighting the constraints and challenges faced by the majority of them and then outline appropriate policy framework for development.

WORLD ECONOMY, THE LDCs AND MINERAL DEVELOPMENT

Any discussion on the world economy, which is basically a global problematique, tends to be vague and often beset with conflicting notions. Since the Second World War, we have witnessed a new form of economic development, a development fostered by deliberate policy and motivated by political will with emphasis on economic interdependence. There has been increasing acceptance of multi-lateral mechanisms of cooperation for the economic well-being of nations. Since 1980, the world economy has been reeling under a severe recession, a Kondratiev cycle of the long wave of economies, and the urgency for development of an actionable blueprint for recovery for some 100 or so developing countries with serious debt-servicing difficulties, has never been greater. If one subscribes to Samuel Johnson, the final quarter of the twentieth century will fall short of the third quarter in its achieved rate of economic progress. Without delving deep into the roots of the current recession which may undermine international economic stability, one must urgently seek a new ideology for resource management, and especially of global mineral resources.

From concerns for the exhaustibility and availability of mineral supplies propounded by the Club of Rome, to the crisis of funding, we have come a full circle when depressed metal prices compounded by falling demand have brought the economic development of many mineral exporting countries almost to a grinding halt. In the immediate foreseeable future, barring disruptions that may occur due to political action in South Africa or elsewhere, growth in real terms of mineral development in most LDCs will be marginal unless new and extraordinary initiatives are taken.

Table 1 summarises the values of a number of basic economic indicators for selected LDCs reflecting the enormous diversity of the nations. For comparative purposes, the values recorded for two DCs (UK and USA) are also shown. Table 2 presents data on the value of mineral production, production per capita and the percentage share of fuels, minerals and metals of the merchandise exports of the nations, which demonstrate forcefully the significant dependence of many of the LDCs on their mineral economy. This is further exemplified by Table 3 which shows the contribution of mining and quarrying to the GNP for selected countries, many of whose fortunes are based on the export of one or two mineral raw materials. While the mineral industry has constituted the cornerstone for general economic development of LDCs with mineral endowment and made significant economic and social impacts, it also has made them vulnerable to depressions in the commodity market and downturns in global economy.

MINERAL DEVELOPMENT IN LDCs - IDENTIFYING THE CONSTRAINTS

Exploitation of mineral resources, be it in a developed or a developing country, is a complex technical activity, often compounded by economic, political and social constraints that tend to make the system totally unwieldy. The unique characteristics of mineral occurrence, the diverse geological and socio-political settings, often in remote or inaccessible areas pose constraints call for specialised inputs that make mining a technology on its own (Ghose, 1985). The chain of mineral development process from an unusual mineral concentration to that of a market demand has many vital links such as infrastructure, skilled human resources and investments. Every mineral project is also marked by significant gestation lag, which aggregatively make mining a high-risk industry. The proportions of costs on infrastructure can be staggering varying from 43.7% to 50.0% of the total project cost as in the case of Cuajone or Ok Tedi, while costs of investment delays can have serious consequences on the viability of projects (Schippers, 1982).

The risks of uncertainity in investment and political climate and the volatility of mineral and commodity prices have constituted of late serious road blocks to investment capital flow into developing countries. Historically, the exploitation of mineral resources in developing countries was effected through transnational mineral enterprises, often leading to oligopolistic controls. The winds of change of nationalism that swept the continents since 1950, with the emergence of independent nations, has transformed the mineral development scene radically. The exercise of sovereign rights over mineral endowment by aspiring LDCs have often led to major disruptions, antagonism and mistrust. The recent past has also witnessed a somewhat subdued role of the transnationals which heretofore have been the classical vehicles of international minerals endeavour. In many developing countries, their role has been assumed by the public sector, which have attained commanding heights in the minerals sector (Ghose, 1985). The performance of the sector, however, has been less than satisfactory in many cases and despite their dominant position, have failed to catalyse the economy. Scathing criticism of the 'millstone' of the public sector and the adverse impacts on national economy have recently been voiced in India and generated a wide-ranging debate (Sathe, 1986). State intervention in the industrialisation process has become widespread in many LDCs and public ownership has been regarded a necessary condition for the establishment of socialism and has been pursued as an end in itself, often without commensurate economic returns.

Quite apart from the above general considerations on the road blocks to mineral development in LDCs, there are also specific issues related to the availability of investment capital, the development of protectionism, the choice of technology and problems of technology transfer (Ghose, 1979). As technologically dependent economies, the LDCs are also faced with the high cost of importing foreign technologies, the loss of national control over decisions regarding technology and often by lack of an effective domestic scientific and innovative capacity. New and emerging forms of resource diplomacy also add a new dimension to the platter of existing problems. In common with the developed world, the mineral industry in developing nations has to contend with the high energy component of operations and consequent impact on operating costs, and the uncertainities of demand and prices in the international market. The widening gulf that has developed between the price levels required to sustain existing operations and that required to encourage new development has been a disconcerting trend of late and the climate for new investment look exceedingly gloomy. Despite warnings voiced by GATT on trade barriers and the continuing efforts by the UNCTAD, the threat of protectionism is very real and holds out the greatest constraint for future mineral development. In the ultimate analysis, however, the overall problems of mineral development cannot skirt the problems of the world economy, whose recovery or relapse will decisively influence the mineral development programmes in LDCs.

AGENDA FOR ACTION - SEARCH FOR CURES AND REMEDIES

As the preceding discussions show, formidable developmental challenges confront the LDCs and indigenous mineral policy initiatives and overall economic reforms apart, some powerful international initiatives are also very necessary. The past two and half decades have seen the substantial contributions of the United Nations Development Fund to mineral resources development in LDCs through systematic exploration and strengthening of national institutions (United Nations, 1979). The existing financial mechanisms such as the World Bank and the Regional Development Banks and new initiatives through the United Nations Revolving Fund for Natural Resources Exploration (UNRFNRE) have significantly aided and supported mineral development efforts. By current reckoning however, the overall level of funding has been woefully inadequate for the magnitude of the challenges. Multi-lateral and bi-lateral aids of a much higher order will be necessary if mineral development is to be stimulated to serve as the engine for growth in the economy of developing countries. A major effort also needs to be launched for establishing international cooperation through a multiplicity of forms, such as the north-south cooperation, south-south cooperation and even regional and sub-regional cooperation.

Developing countries must also adopt some pragmatic approach towards mineral policy for development, which while based on increasing self-reliance or collective self-reliance, must also attract and encourage capital from outside, develop indigenous capability and assure greater measure of regional cooperation. What is needed therefore is a holistic mineral policy appropriate to each developing nation's situation to stimulate mineral development. Likewise, strong political commitment, flexible implementation of policies and sustained efforts - hallmarks of capable management - can help attain the mineral policy goals.

There are also some basic economic issues that need to be addressed on a global basis to set the terms for a new development strategy. Industrial countries provide a market for about 65 per cent of the developing country's export. Their buoyancy - or lack of it - and the amount of trade protection they choose to employ will weight heavily on the developing countries. The occurrence of minerals, as has rightly been pointed, is an accident of nature and its contribution to economic growth depends on the knowledge of the choices and the wisdom of policies (Gluschke, 1982).

CONCLUDING REMARKS

With the global economy in doldrums, the mineral development programmes in many developing countries face crises of unprecedented magnitude. Investment in new mine capacity has come to a virtual standstill and mine closures have taken place on a scale that has not been seen since 1930s. In such an international climate for mineral industries, two distinct scenarios of development could be projected. The first scenario is one of a multipolar world with rampant protectionism, cartels and lack of accomodation with serious strains on the economy of the developing nations, which may aggravate political instability. In the other scenario, the developed world increases its quantum of development assistance and a new dimension in international cooperation is reached to help catalyse the global economy for the most effective and direct results on mineral development. Whatever be the scenario, the LDCs in their sectoral plans must pursue aggressively a multiplicity of goals to search for and develop new mineral deposits, find new sources of finance, increase export earnings, diversify mineral production, and improve physical infrastructure. While the specific policy agenda for each developing country will depend on its political culture, the nature of problems it faces and what it has already achieved, the overriding policy concern must focus on productivity, employment generation and environmental protection for the health of the mining industry. The gravity of the global problematique vis-a-vis mineral development has been realised. The next step hopefully is to move up the learning curve from 'concern' to 'solution'.

The marked growth in the mineral resource base in developing countries implies that the focal point of mining activities will in the future be concentrated in LDCs. The world, which is passing through a critical phase in its history, has to test its ability to cooperate in developing the minerals and energy sectors on a truly global basis.

REFERENCES

Gluschke, W., Iwase, N., and Zorn, S., 1980: Mineral Resources and the Impact of Mining on LDCs, in Mining for Development in the Third World (Ed: S. Sideri and S. Johns), Pergamon Press.

Ghose, A.K., 1985: 12th World Mining Congress - An Overview in Digest of the 12th World Mining Congress, 1-4.

Schippers, H.J., 1982: Mining and Infrastructure in Developing Countries, Natural Resources and Development, Vol. 16, 27-39.

Ghose, A.K., 1985: Mineral Resources Development in India - Role of the Public Sector, in Role of Government in Mineral Resources Development, Institution of Mining & Metallurgy, 132-135.

Sathe, V., 1986: Articles in Times of India, 5th, 6th and 7th August.

Ghose, A.K. and Malhotra, M.L. (Ed), 1979: Transfer of Technology in Mineral Industries in Developing Countries, Institution of Engineers(India).

United Nations, 1981: Two Decades in Mineral Resources Development, in National and International Management of Resources (Ed: M. Jones), Institution of Mining & Metallurgy, 163-177.

Gluschke, W., 1982: Third World Mineral Development: Policies and Structures, in Legal Institutional Arrangements in Mineral Development, Mining Journal Books, 3-10.

TABLE 1

Selected Economic Indicators for 32 LDCs and 2 DCs

Country	Population (in millions) (mid-1983)	Density of population (per sq. km)	GNP/capita Dollars (1983)	Av. Annual Growth rate (1965-83)
Low income economies				
India	733.2	223	260	1.5
Tanzania	20.8	22	240	0.9
China	1019.1	107	300	4.4
Sri Lanka	15.4	233	330	2.9
Pakistan	89.7	112	390	2.5
Ghana	12.8	54	310	-2.1
Haiti	5.3	189	300	1.1
Madagascar	9.5	16	310	-1.2
Kenya	18.9	32	340	2.3
Afghanistan	17.2	27	-	0.5
Lower middle income economies				
Lesotho	1.5	50	460	6.3
Liberia	2.1	19	480	0.8
Bolivia	6	5	510	0.6
Indonesia	155.7	81	560	5
Zambia	6.3	8	580	-1.3
Egypt	45.2	45	700	4.2
Zimbabwe	7.9	20	740	1.5
Morocco	20.8	46	760	2.9
Peru	17.9	14	1040	0.1
Congo	1.8	5	1230	3.5
Jamaica	2.3	209	1300	-0.5
Ecqudor	8.2	28	1420	4.6
Colombia	27.5	24	1430	3.2

contd....

Contd.

Country	Population (in millions) (mid-1983)	Density of population (per sq.km)	GNP/capita Dollars (1983)	Av.Annual Growth rate (1965-83)
Upper middle income economies				
Malaysia	14.9	45	1860	4 . 5
Chile	11.7	15	1870	-0 . 1
Brazil	129.7	15	1880	5
Mexico	75	38	2240	3 . 2
Algeria	20.6	9	2320	3 . 6
South Africa	31.5	26	2490	1 . 6
Yugoslovia	22.8	89	2570	4 . 7
Venezerela	17.3	19	3840	1 . 5
Industrial Market economies				
U.K.	56.3	230	9200	1 . 7
U.S.A.	234.5	25	14110	1 . 7

Source: World Development Report 1985, Oxford University Press

TABLE 2

Value of Mineral production, production per capita and share of fuels, minerals and metals in merchandise export

Country	Value of minerals (1978) (million Dollars)	Mineral production per capita in Dollars	Percentage share of fuels, minerals and metals in merchandise exports (1982)
Low income economies			
India	2330.5	3.2	7
Tanzania	12.9	0.6	5
China	31812.6	31.2	26
Sri Lanka	7.7	0.5	14
Pakistan	312.4	3.5	6
Ghana	119.3	9.3	NA
Haiti	11	2.1	NA
Modagascar	17.6	1.9	12
Kenya	17.6	0.93	29
Afghanistan	116.9	6.8	NA
Lower middle income economies			
Lesotho	2.2	1.5	NA
Liberia	357.1	170	67
Bolivia	688.9	115	93
Indonesia	9034.9	58	85
Zambia	782.9	124.3	97
Egypt	2374.6	52.5	70
Zimbabwe	402.6	50.9	24
Morocco	707	34	39
Peru	1801	100.6	69
Congo	226.8	126	
Jamaica	193	84	22
Ecqudor	938.1	114	64
Colombia	902.7	32.8	8

contd....

Contd.

Country	Value of minerals (1978) (million Dollars)	Mineral production per capita in Dollars	Percentage share of fuels, minerals and metals in merchandise exports (1982)
Upper middle income economies			
Malaysia	1865.3	125.2	35
Chile	1835.9	157	65
Brazil	2310,3	17.8	18
Mexico	7371.2	98.3	78
Algeria	6810.4	330.6	99
South Africa	8107.7	257.4	14
Yugoslovia	1175.9	51.6	6
Venezerela	103771	5998	97
Industrial Market economies			
U.K.	12350.4	219.4	24
U.S.A.	73837.7	315	9

Source: 1. World Development Report 1985, Oxford University Press.

2. F. Callot: World Production and Consumption of Minerals in 1978. Mining Journal Books Ltd., 1981

TABLE 3

Contribution of Mining and Quarrying to GNP in Selected Countries

Country	Year	Share of GNP(GDP) in Mining & Quarrying(%)	Major Products
Bolivia	1984	6	Tin and antimony
Mexico	1984	1.3	Copper, Lead, Zinc, Silver
Brazil	1884	3.85	Iron ore, Gold
Chile			
Peru	1984	9.6	Copper, lead, zinc, silver
Guyana	1983	16	Bauxite
Surinam	1983	15	Bauxite
India	1985	1.4	Coal, iron ore, manganese
Botswana	1984	45	Diamonds
Namibia	1984	28	Diamonds
Swaziland	1982	5	Asbestos

ROLE OF GOVERNMENT IN MINERAL DEVELOPMENT IN INDIA

S.V.Ali
Mineral Development Board
New Delhi, India.

ABSTRACT

Production of minerals has witnessed significant increase since the coming of Independence in India. Side by side, considerable inventory build-up has also taken place. This is necessarily the outcome of efforts put in by the Government of India and its agencies. The planning mechanism in the mineral development gets into motion at the Planning Commission providing a co-ordinating umbrella to the central ministries and State planning bodies. The exploration inputs are provided by the Geological Survey of India, Mineral Exploration Corporation and State Directorates of Geology and Mining. The regulation aspect is looked after by the Department of Mines and the State Governments. Indian Bureau of Mines is responsible for the scientific development and conservation of mineral deposits. The Department of Environment takes care of environment vis-a-vis mining. R&D resource inputs are provided by the Central Mining Research Station and Mineral Development Board besides concerned ministries. The new field of nodule mining is on the agenda of the Department of Ocean Development. Mineral beneficiation requirement is looked after by the Indian Bureau of Mines, Regional Research Laboratory, Bhubaneswar, etc. The safety in mines is the responsibility of the Directorate General of Mines Safety.The allocation of funds for mineral development is an exercise undertaken by the Planning Commission. Government of India also negotiates with outside agencies and Governments for resource pooling in technology and finances.

INTRODUCTION

Though mining was perhaps born the day the primitive man picked up a piece of flint, it is the coming of Independence which catalysed the rightful thrust for mineral development in India. The story of building up the consumption of coal from an insignificant quantity of 20,000 t per year in the early nineteenth century to about 30 mil t in the late forties is fascinating indeed which could not, however, predicate the rapid strides which followed taking it to five times as much by mid-eighties.

The value of mineral production since 1947 till 1984 went up from Rs.64 crores to Rs.7680 crores which works out to a near-quantum jumps of 120 times (Table I). This figure was 90 times since 1951 (i.e. the year of introducing the concept of planned development in the country) and 3.33 since the beginning of the present decade. Some experts might not reconcile with this type of comparison for growth identification on account of the monetary devaluation and escàlation which took place in the meantime. Reverting therefore, to quantitative appreciation of the situation, it is seen that the overall production increase since 1947 till the end of the 6th Five Year Plan has been 6.5 times and since the 1st Plan by about 6 times. During the period 1951 to 1985, the growth in mineral-wise production has been as follows (Table 2).

Coal	4.2 times
Iron ore	13.2
Limestone	16
Dolomite	228
Bauxite	29.8
Gypsum	5.8
Chromite	27

Side by side with the production growth, inventory build up was also witnessed (Table 3). Discoveries in respect of bauxite reserves could be described as the most significant one. During the period 1960-61 and as one date, the inventory of bauxite has gone up by 35.8 times from 74 to 2653 mil.t. During the same period, the reserves of coal and iron ore were augmented by2.7 times each, copper by 17.5 times, lead-zinc by 39.8 times and dolomite by 13.1 times.

ROLE OF GOVERNMENT

In the above growth scenario, the Government of India has been playing a consistent and effective result-oriented role of coordinating arm for planning, resource pooling, monitoring, regulation, etc. Let us review each of the above aspects in detail.

Planning

With the coming of political Independence, the concept of planned development was introduced in the country with the object of building a self-reliant economy for generating a social system based on equality and justice. The goal of growth of the economy warranted growth in industrial scenario which had its own snowballing effect in the consumption pattern of minerals. Starting from commencement of the First Five Year Plan in 1951 to the onset of the 7th Plan efforts for boosting mineral development activities in the country were consistently on the agenda for action recognising the non-renewable nature of mineral deposits and need for regulation through scientific development and conservation. Industrial Policy Resolution (1956) provided an effective guideline which delineates the industries for State and private sectors.

Planning in India is the domain of the Planning Commission at the Centre. A comprehensive exercise carried out every five years with the assistance of expert groups, concerned ministries in the Statess and the Centre, enables it to work out growth rates, production-consumption targets,resource availability and allocation. The 7th Five Year Plan (1984-85 to 1989-90) which was approved by the National Development Council on 9.11.1985, identified a growth rate of a little over 5 per cent,, and laid emphasis on modernisation of industry, introduction of high technology, achieving reduction in cost, improvement in productivity, human resource development, augmentation in infrastructure, improving work ethics, preserving environment and ironing out regional disparities. A perspective look over a period of 10 years beyond the 7th Plan and upto 2000 A.D. is a special feature of the 7th Plan document.

In planning for mineral development during the 7th Plan, the above policy options have been kept in view. The annual compound growth rate since the onset of the 7th Plan to the year of its oulmination is 14.9 per cent for lignite followed by 13.7 per cent for lead, 12.5 per cent for aluminium, 10.2 per cent for cement, 9.1 per cent for zinc and 8.9 per cent for coal. Growth rates in respect of iron ore, copper and steel have been fixed at much levels (Table 4).

Looking at the targets set by the Planning Commission for the year 2000 A.D. (Table 5), it is seen that a growth of 3.04 times has been envisaged in respect of refined copper in ten years beyond the 7th Plan. Taking into account that not much of growth was planned for this metal during the 7th Plan, it can be concluded that the planners are heavily counting on the expansion envisaged for the Malanjkhand copper project. In respect of other nonferrous metals, the growth for lead has been set at 2.22 times, for aluminium at 1.70 times and for zinc at 1.68 times. For fuel minerals, the figure is 1.85 for coal and 1.97 for lignite. The growth profile for steel and iron ore will somewhat subdued at 1.66 and 1.46 times during this period.

In order to achieve the targets, the monetary resource input allocation is also carried out by the Planing Commission in consultation with the Finance Ministry and other concerned ministries. It is seen that of the total plan allocation of Rs.95,534 crores in the central sector, 10.7 per cent has been earmarked for mineral development activities (Table 6). As to the State sector and Union Territories, the percentage share for mineral development has been fixed at 0.39 and 0.24 per cent respectively. The obvious reason for such allocation is the fact that most of major mineral development activities in the country are confined to the central sector. Coal (Rs.7,400 crores) and non-ferrous minerals (Rs.1,785 crores) together have the lion's share at 89.6 per cent of the total allocation for the central sector.

Looking at the mineral development programmes in the State sector, it is seen that of the total outlay of Rs.321.40 crores, sizeable amounts are scheduled to be shared by Andhra Pradesh (Rs.56.70 crores), Bihar (Rs.56.0 crores) and Rajasthan (Rs.78.40 crores). Amongst the Union Territories, Delhi has been allocated as much as Rs.8.0 crores of the total Rs.9.0 crores (Table 7).

Exploration

In the successive Five Year Plans, the institutional framework for mineral exploration received much support. The Geological Survey of India was expanded and equipped with modern facilities, Mineral Exploration Corporation was created, and the Directorates of Geology and Mining of various States were created and/or strengthened. During the second Five Year Plan, emphasis was laid on the exploration for coal and iron ore and other minerals required for steel and base metals. During the Third Plan, priority was assigned to base metals, bauxite and phosphorite horizons. During the Fourth and Fifth Plans, coal exploration was again on priority. The need to establish reserves of steel making minerals other than iron ore was also felt. Additional reserves of manganese ore, chromite and refractory raw materials were established. This was also the period when the work on East Coast bauxite bore fruits which brought India in the forefront of bauxite resources from the position of being perilously deficient. The Sixth Plan witnessed substantial increase in exploration activities for coal, base metals, manganese ore, chromite, phosphorite, potash, etc. Work was also taken up on enlarged scale to establish strategic minerals such as tin and tungsten. Significant mineral finds during the 6th Plan included Rampura-Agucha lead-zinc deposit in Rajasthan, Chicargunta gold deposit in Karnataka and Tesham tin deposit in Haryana.

Planning Commission has allocated Rs.125 crores for Mineral Exploration Corporation and Rs.95 crores for Geological Survey of India for the 7th Plan. During this Plan, emphasis on modernisation is on the anvil. Photo Geology/remote sensing, non-coring drilling, oil well drilling, multi probe ggeophysical borehole logging, electrical resistivity surveys, seismic refraction surveys, high resolution seismic surveys, physico-geological investigations, optimisation of exploration inputs, development of software for processing voluminous exploration data, etc. would assigned increasing importance. Coal exploration todeeper horizons would be taken up in a big way. Exploration for ferro-manganese nodules in the deep ocean would be further taken up.

Planning Commission has calculated the critical life for the recoverable reserves as on 1.1.1995 after depleting the production for the period 1980-94 from the recoverable reserves as on 1.1.1080, at the frozen rate of production as at 1994-95 level (Table 8). It is seen that the balance life in respect of copper ore, kyanite, lead, manganese ore, rock phosphate, sillimanite (massive variety) has already reached critical levels and special efforts are warranted during the coming Plans to orient activities for augmenting reserves.

In order to optimise the utilisation of resources allocated and for avoiding duplication and overlapping of efforts, creation of an institutional framework is warranted to formulate and co-ordinate the programmes of survey, exploration and exploitation of minerals.

Regulation

The regulation aspect of mineral development is looked after both at Central and State levels.

The draft of the National Mineral Policy is understood to have been circulated to the appraising organisations of the Government only recently. It would cover all minerals other than coal, lignite and petroleum. The object of the National Mineral Policy is to ensure adequate supply of mineral based material inputs for industry, optimal use of available mineral resources, promotion of necessary backward and forward linkages, encouragement of R&D and minimising the adverse efforts of mineral development on environment through appropriate measures, development of infrastructure and promotion of regional development. Utilization of lower grade minerals and ores, introduction and adoption of latest agglomeration and beneficiation techniques, optimal mechanisation and productivity improvement are the other significant features of the draft policy. Appropriate changes would, however, be required from time to time tailoring it to the changing mineral scenario and its economics.

A bill on the amendment of Mines and Minerals (Regulation and Development) Act has been recently introduced in the Parliament, which proposes toadd 11 more minerals in the First Schedule of the Act (Table 9), raising punishment for illegal mining from 6 months to one year and fine from Rs.1000 to Rs.5000, measures to removing certain bottlenecks by dispensing with certificate of approval, income tax clearance certificate for prospecting, premature termination of prospecting licenses or mining lease on ecological or other grounds, preparation of a mining plan as a precondition for the grant of a mining lease, rationalising the period of mining lease and their renewal, fixing periodicity for royalty revision and dead rent.

Indian Bureau of Mines, under the Department of Mines, has been playing a pivotal role in respect of mineral development, scientific exploitation and conservation. This has been possible through their inspecting teams and regional offices located at all mining centres such as Ajmer, Bangalore, Calcutta, Dehradun, Goa, Hazaribagh, Hyderabad, Jabalpur, Madras, Nagpur, Nellore and Udaipur. Their enormous data bank also has been serving the mining industry as a useful tool for planning and monitoring.

There are about 3000 small mines in India of which over 800 are in Rajasthan alone (Table 10). Gujarat comes next with about 450 mines. Of late, the significance of small scale mining has been realised in India as about 50 per cent of mineral contribution comes from the same. There is a need to define small scale mining since they are in need of appropriate support from the Government on account of limited technical, financial and physical resources available to them. The vigorous efforts initiated some time back for defining smallscale mining activity are still far from a rriving at any logical conclusion.

Environment

Indiscriminate mining activities resulted inenvironmental ravages with distressing examples at Jharia, Dehradun, Goa, etc. The delicate balance between mining activities and environmental preservation seems to be currently tilting towards the latter. The difficult situation has been realised on account of the fact that the country has 15 per cent of the world population as against 2.5 per cent land availability. This exerts tremendous pressure on land, the availability of which will reduce to one third at 0.33 hectare by the year 2000 A.D. from the level of 1951.

Efforts are being made not only to contain the damages alreayd done to environment but to improve the existing situation. A new Department of Environment has been created. The environment aspect has also been included in the Charter of Functions of the Indian Bureau of Mines. Environment protection plan has become an integral part of all mining projects. Dhanbad Declaration of March 1982 has also rendered singular service by calling for the integration of environmental dimensions in the planning, design, development and operation of mines. Introduction of environmental element at the conceptual stage of mining, incorporation of environmental science in educational syllabi, creation of generalawareness, formation of practical codes, taking up afforestation, soil conservation - land reclamation programmes, water, smoke, air and dust surveys, measuring noise and vibration levels, introduction of monitoring of remedial programme, creation of effective monitoring system, etc. are the significant steps in the right direction. The Bill seeking to amend the Mines and Minerals (Regulation and Development) Act also provides for termination of prospecting license or mining lease on ecological grounds.

R&D efforts

The need for R&D efforts in mining has been realised in the correct perspective. Establishment of organisations such as the Central Mining Research Stationand Mineral Development Board (formerly Iron Ore Board) was the outcome of this policy. During the 7th Plan, Aluminium Research, Design and Development Centre is likely to come up for which a provision of Rs.10 crores has been made. The organisation will serve to promote the use of aluminium metal in view of sizeable production capacity.

There are S&T cells in concerned Ministries to monitor, and to avoid overlapping, duplication, infructuous work and promote co-ordination in R&D work. They also work to identify the thrust areas warranting R&D inputs. Once this is done, the work is entrusted to the laboratories or R&D cells of central undertakings. Thus these cells are no more regarded as mere appendage of fashion but have become extremely useful acquirement. The ore beneficiation laboratories of Indian Bureau of Mines at Nagpur, Ajmer and Bangalore, and Regional Research Laboratory, Bhubaneswar are known for their valuable work in the concerned field. The Department of Ocean Development has been promoting the pioneering work in ferro manganese nodule exploration and mining from the deep sea. The work on extraction of metallic values from the nodules is under way at Indian Bureau of Mines, Hindustan Copper Ltd. and Hindustan Zinc Ltd.

Safety

Safety in mines is the responsibility of the Directorate General of Mines Safety. This important organisation has been playing on extremely effective role with its headquarters at Dhanbad and zonal and regional offices at all the points of mining activities ksuch as Dhanbad, Sitarampr, Nagpur, Hyderabad, Ranchi, Koderma, Chaibasa, Ajmer, Udaipur, Jabalpur, Bhubaneswar, Goa, Oorgaum, Digboi, Bermo, Ramgarh, Gaziabad, Manendragarh, Parasia, Ramagundam, Nellore, Bellary, etc. The national mine safety record with low accident rate is to the credit of DGMS.

Monitoring

Hitherto, the monitoring of projects under construction was the responsibility of the conceerned Ministries and the Planning Commission. A new Ministry for Programme Implementation has now been created which monitors the progress of all the major projects in the country. The Ministry identifies the elements of time and cost over-runs and also the reasons for the same. Only recently an 'ultra' expert group with top experts from the private sector has been constituted by the Ministry to suggest ways and means to contain the problem of time and cost over runs in project implementation.

Support from outside

agreement with regard to large projects in order to secure technical, barter and financial support. The project of National Aluminium Co.Ltd. now costing over Rs.2200 crores, comprising of a 2.4 mil.t.p.a. bauxite mine, 800,000 t.p.a. alumina plant and 218,000 t.p.a. of aluminium smelter is an outcome of such policy. Rampura-Agucha project kbased on a 60 mil.t. of lead-zinc deposit having as much as 15 per cent metal content and with imperial process smelter of annual capacity 70,000 t of zinc and 35,000 t of lead requiring a total funding exceeding Rs.600 crores, would be another project seeking such support. Expansion programme of Malanjkhand copper project taking its present annual capacity of 2 mil.t. ore to about 10 mil.t. of ore in view of the recent augmentation of reserves exceeding 300 mil.t., would also attract similar approach.

Jharia Master Plan which has been drafted by Indian and Polish experts as a result of deliberations over a period of 6 years aiming to retrieve locked up 2,000 mil.t. of primary coking coal, envisages the construction of 21 underground and 9 open cast mines and would be a fit case for this type of assistance.

Negotiations with large number of countries and various bodies are carried out for resource pooling in technology and finances. The following are the examples from just one country - USSR:

1. Preparation of feasibility reports for Mukunda (12 mil. t.p.a.), Nigahi (14 mil.t.p.a.), Kumari (10 mil.t.p.a.), Mohar (10 mil.t.p.a.), etc.

2. Development of coal mines in Godavari basin for Singareni Collieries Co.Ltd.

3. Working Tipong mine in Makum Coal Field.

4. Negotiating the Andhra Alumina project.

5. Work relating to other schemes such as geological studies, exchange of information on new technology on coal enrichment, coal gasification, etc.

EPILOGUE

From the above, it is obvious that the role played by the Government and its agencies has been very effective in India's mineral development efforts. The significant and effective role will continue on account of the fact that the mineral development is commanding increasing attention in a developing economy like ours. Support of the Government is decided basing on the facts of national priorities. Kudremukh iron ore project costing over Rs.600 crores is a case in point. This project designed to produce 22.5 mil.t. per year of ROM and 7.5 mil.t. per annum of concentrates could not perform well on account of the failure of Iran to import the agreed quantities. A pellet plant was erected at Mangalore at a cost of more than Rs.100 crores to utilise

3 mil.t. of concentrates per year to salvage Kudremukh project. It was again the Government which sponsored negotiations for promoting the exports of these concentrates. The result of such policy was the successful negotiation with Japan agreeing to import 25 mil.t. of concentrates over a period of 10 years.

On the other hand, mining units performing well also receive encouragement from the Government. Neyveli Lignite Corporation provide a typical example in this respect. By the year 2000 A.D., the lignite production capacity of this complex would be raised to 32 mil.t. per annum requiring 400 mil.t. of excavation in overburden alone. During the 7th Plan itself, the Government has allocated Rs.700 crores for lignite sector alone.

TABLE - 1

Growth of Mineral Production in India

Year	Total value of mineral production (Rs. crores)*
1947	64.0
1951	85.2
1961	176.2
1971	503.4
1975	1,111.9
1980	2,304.8
1983	6,796.6
1984	7,680.0

N.B.: * 1 crore = 10 million

TABLE - 2

PRODUCTION OF MINERALS

(Million Tonnes)

Minerals/ Ores	Years 1951	1965	1975	1980	1981	1982	1983	1984	1985
Coal	35.0	67.16	95.91	109.2	123.1	128.5	136.3	144.5	147.0
Iron Ore	2.27	23.87	41.79	41.90	41.40	42.0	38.7	41.9	43.12
Limestone	2.95	19.9	26.5	29.0	32.0	33.0	38.0	45.0	47.22
Dolomite	0.01	0.98	1.46	2.03	2.07	2.20	2.20	2.30	2.28
Bauxite	0.07	0.71	1.27	1.78	1.95	1.95	1.93	2.03	2.09
Gypsum	0.21	1.16	0.82	0.87	0.96	0.96	0.90	1.20	1.21
Chromite	0.02	0.06	0.50	0.32	0.34	0.36	0.44	0.45	0.54

TABLE - 3

MINERAL INVENTORY

		Reserves(Mil.t)	
	Minerals/Ores	As on 1960-61	As on date
1.	Bauxite	74	2,653
2.	Copper	33	578
3.	Lead-zinc Ore	9	358
4.	Iron Ore	6,400	17,565
5.	Manganese Ore	100	135
6.	Rock Phosphate	-	120
7.	Dolomite	301	3,947
8.	Limestone	-	73,199
9.	Coal	41,000	112,000

TABLE - 4

COMPOUND ANNUAL GROWTH RATES FROM 6TH TO 7TH PLAN

Sr.No.	Commodities		Year		Compound Annual Growth rate
1	Coal	mil.t	147.44	226	8.9
2.	Lignite	mil.t	7.8	15.2	14.9
3.	Iron Ore(1)	mil.t	42.2	58.1	6.6
4.	Cement	mil.t	30.1	49	10.2
5.	Steel(2)	mil.t	8.77	12.64	7.6
6.	Aluminium	'000t	276.5	499	12.5
7.	Copper(3)	'000t	33.5	42.7	5.0
8.	Zinc	'000t	57.6	89	9.1
9.	Lead	'000t	14.2	27	13.7

(1) Concentrates are included

(2) Saleable

(3) Refined

TABLE - 5

OUTPUT PROJECTIONS

Sr.No.	Commodities		Years	
			1989-90-	1999-2000
1.	Coal	mil.t	226	417
2.	Lignite	mil.t	15.2	30
3.	Iron Ore(1)	mil.t	49	87
4	Cement	mil.t.	49	87
5.	Steel(2)	mil.t.	12.64	21
6.	Aluminium	'000t.	499	850
7.	Copper(3)	'000t	42.7	130
8.	Zinc	'000t.	89	150
9.	Lead	'000t.	27	60

(1) Concentrates are included

(2) Saleable

(3) Refined

TABLE - 6

7TH PLAN OUTLAY

			(Rs.Crores)
I.	Public Sector	1,80,000	
	Central		95,534
	State		80,698
	Union Territories		3,768
II.	Private Sector	1,40,000	
	Total	3,20,000	
	Outlay for mining sector		
	Central	10,245.58	
	State	321.40	
	Union Territories	9.0	
I.	Department of Mines		1,050.00
i)	Bharat Aluminium Co.Ltd.,		331.0
ii)	National Aluminium Co.Ltd.		1,174.0
iii)	Hindustan Copper Ltd.		165.0
iv)	Hindustan Zinc Ltd		104.0
v)	Bharat Gold Mines Ltd.		21.10
vi)	Mineral Exploration Corporation		125.0
vii)	Geological Survey of India		95.0
viii)	Indian Bureau of Mines		14.0
ix)	a) Aluminium Research,Design and Development Centre.		10.0
	b) S & T Schemes		9.9
x)	Andhra Alumina Project		1.0
II.	Department of Steel		6,420.3
	Ferrous Group of Minerals		200.0
	i) National Mineral Development Corporation		145.30

ii) Kudremukh Iron Ore Co.Ltd.	18.45
iii) Manganese Ore (India) Ltd.	18.80
iv) Mineral Development Board	5.0
v) Loan for water supply	12.45
III. Department of Coal	7,400.58
i) Coal India Ltd.,	6,000.58
ii) Singareni Collieries Co.Ltd.,	580.0
iii) Neyveli Lignite Corporation	700.00
iv) R&D	120.0
IV. Department of Fertilizers	2,025.75
Pyrites Phosphates and Chemicals Ltd.	47.0
V. Department of Atomic Energy	1,075.0
i) Atomic Minerals Division	5.0
ii) Indian Rare Earths Ltd.	30.0
iii)Uranium Corporation of India Ltd.	90.0
VI. Department of Ocean Development	10.0
VII. Department of Public Enterprises	1,654.80
Cement Corporation of India	413.0

TABLE - 7

STATEWISE OUTLAYS FOR MINERAL DEVELOPMENT

S.No.	States	7th Plan outlay (Rs.Crores)	
1.	Andhra Pradesh	56.70	
2.	Assam	4.25	
3.	Bihar	56.00	
4.	Gujarat	18.67	
5.	Haryana	0.70	
6.	Himachal Pradesh	1.50	
7.	Jammu & Kashmir	3.50	
8.	Karnataka	5.00	
9.	Kerala	4.00	
10.	Madhya Pradesh	9.98	
11.	Maharashtra	3.10	
12.	Manipur	0.60	
13.	Meghalaya	1.80	
14.	Nagaland	8.70	
15.	Orissa	20.00	
16.	Punjab	0.30	
17.	Rajasthan	78.40	
18.	Sikkim	1.50	
19.	Tamil Nadu	10.00	
20.	Tripura	0.10	
21.	Uttar Pradesh	28.60	
22.	West Bengal	8.00	
	Sub Total		321.40
	Union Territories		
1.	Andaman & Nicobar	-	
2.	Arunachal Pradesh	0.50	

3.	Chandigarh	-
4.	Dadra & Nagar Haveli	-
5.	Delhi	8.00
6.	Goa, Daman & Diu	0.30
7.	Lakshdweep	0.05
8.	Mizoram	0.15
9.	Pondicherry	-
	Sub Total	9.00
	Total	330.40

TABLE - 8

LIFE INDICES OF CRITICAL MINERALS

Sr. No.	Minerals	Recoverable Reserves as on 1.1.80 (mil.t)	Depletion during 1980-94 (mil.t)	Recoverable reserves as on 1.1.1995 (mil.t)	Projected Production during 1994-95(mil.t)	Balance life sy 1994-95 level of production (Years)
1.	Copper					
	i) Only producing mines	1.95	0.78	1.17	0.065	18
	ii) All economic and marginally economic deposits.	3.15	0.78	2.37	0.065	36
2.	Kyanite	1.43	0.65	0.78	0.054	14
3.	Lead					
	i) Only producing mines	1.687	0.522	1.165	0.059	20
	ii) All deposits	2.746	0.522	2.234	0.059	38
4.	Manganese Ore					
	i) High grade	20	8	12	0.79	15
	ii) Low grade	54	13	41	1.34	31
5.	Rock phosphate only high grade	20	12	8	1.1	8
6.	Sillimanite (massive)	0.213	0.116	0.097	0.013	7

TABLE - 9

LIST OF MINERALS UNDER THE AMENDED FIRST SCHEDULE

Apatite and phosphatic ores, barytes, bauxite, beryl and other beryllium bearing minerals, chrome ore, coal and lignite, copper ore, dolomite, gold, gypsum, iron ore, kyanite, lead, limestone, lithium bearing minerals, magnesite, manganese ore, minerals of 'rare earths' (containing uranium, thorium), molybdenum, niobium bearing minerals, nickel ores, phosphorite and other phosphatic minerals containing uranium, platinum and other precious metals and their ores, precious stones, rutile, silver, sillimanite, sulphur and its ores, tantalium bearing minerals, tin, tungsten ores, uraniferous allanite, monazite and other thorium minerals, uranium bearing tailings left over from ores after extraction of copper and gold, ilmenite and other titanium ores, vanadium ores, zinc and zircon.

TABLE - 10

NUMBER OF SMALL MINES IN INDIA

State/Union Territories	Number	
Andhra Pradesh	313	
Bihar	224	
Gujarat	447	
Haryana	8	
Himachal Pradesh	18	
Jammu & Kashmir	3	
Karnataka	223	
Kerala	28	
Madhya Pradesh	322	
Maharashtra	82	
Manipur	2	
Orissa	112	
Rajasthan	815	
Tamil Nadu	120	
Uttar Pradesh	103	
West Bengal	18	
Delhi	3	
Goa	116	
Total		2962

ROLE OF THE GOVERNMENT IN RELATION TO MINERAL EXPLORATION AND DEVELOPMENT

S A Bilgrami
Resource Development Corporation (Pvt) Ltd
Karachi, Pakistan

ABSTRACT

The common problems faced by mineral industry in less developed and developing countries are lack of know-how, evaluation facilities, financial resources and incentives, infrastructure and excessive government control, high cost of energy, etc. It has been suggested that the quality of geoscience education and mineral deposit evaluation facilities should be improved by setting up pilot plants and enhancing laboratory facilities. In evaluating mineral projects due emphasis should be laid on socio-economic benefits of such projects most of which are located in isolated areas. Governmental support for developing infrastructure, encouragement for mineral-based industries, low railway freight, liberal import of mining/mineral processing equipment which should be exempt from import duties and sale taxes and special low interest financing are some of the basic incentives needed to develop a viable mineral industry.

INTRODUCTION

In most developing countries, the industry is under rigid control of the government. Not only has the investor to seek government approval for setting up an industry, he has also to depend upon the government for permits, licences, power, gas, water and railway connections. In addition, in some countries the area where a particular industry may or may not be established is delineated by the government. Thus the industrialist is at times guided by government directives rather than economic considerations. In outlining the role of government in mineral exploration and development it will be pertinent to first recount briefly the problems of the mineral industry so that the role of government can be more easily identified.

CHARACTERISTIC PROBLEMS OF THE MINERAL INDUSTRY

Lack of knowhow

Mineral industry uses a wider range of technical expertise than any other industry. From unskilled labour to most sophisticated engineering skills are needed to locate, evaluate, mine and process minerals. Usually fairly good geological expertise is available within most of the developing countries but there is a wide gap in evaluation and processing areas. This is made more serious by the absence of suitable laboratory facilities.

Lack of Evaluation Facilities

Complete evaluation of mineral projects needs not only expertise in geology and mining but also in mineral dressing and pyro and hydrometallurgy. In many cases qualified persons are available but the equipment necessary for a complete evaluation is lacking. Facilities for geotechnical studies, pilot plant tests and monitoring of the impact on environment are lacking. Feasibility studies, based on bench scale tests, are not reliable and, therefore, not acceptable to investors and financiers.

Lack of financial resources

Mining is a high risk industry. This, coupled with the long gestation period, makes it unattractive for investment. The occurrence of mineral deposits in isolated and relatively undeveloped areas not only increases the cost of mineral development but also makes it unattractive for skilled people. Lack of expertise for economic evaluation of mineral projects in government planning departments as well as in financial institutions compounds this difficulty thereby making it doubly difficult to raise finances for mineral projects. Wide fluctuations in metal prices make a long-range financial evaluation of metal mining projects more difficult. Hence, financial institutions wish to avoid such projects since they have more attractive avenues for investment where results are usally predictable with greater degree of accuracy.

Mineral projects are radically different from other industrial projects say textile mills, fertilizer plants etc. Thus when mineral projects are judged by the same yardstick as other industrial projects, they fail to come up to the standards of the evaluatig agency - usually government planning department or a bank. Governments in Less Developed Countries (LDCs) encourage more labour intensive projects but mineral processing plants now are fully automated and hence do not create as many jobs per say one million dollars as other industries, hence are considered less attractive from this point of view.

It is not generally appreciated that mineral projects initiate a chain of development which is long and all encompassing. Job opportunities created by processing industry in support industries are usually not taken into consideration. Mineral deposits are often located in undeveloped, isolated areas where new townships come up leading to overall development of the area.

Laws governing the Mineral Industry

Mineral industry has to deal with at least six or seven different ministries in the government and several other departments. This makes it difficult for the industry to get quick decisions. A considerable time has to be spent in filling of forms, submission of returns/statements which are not even read and do not really serve any purpose.

High Cost of Energy

Mineral industry is energy intensive and the cost of electric power is usually very high in most developing countries. Gas, a comparatively cheaper fuel, is usually not available. Suitable coal, if it has to be imported, is expensive and necessary foreign exchange may not be available. This puts a restraint on processing of ores domestically.

Lack of Engineering Industry and Fabrication Facilities

Mining industry uses a wide range of electrical and mechanical equipment. From simple pick and hammer to most advanced electronically operated and controlled machines are used. A well developed engineering industry is desirable to support the mining industry not only for day to day operations but also for help in fabrication of equipment and modifications of readily available plant and machinery. Lack of fabricating facilities also creates serious problems in case of breakdowns which cannot be quiuckly repaired. Sometimes, necessary spare parts are not readily available and cannot be imported in good time due to foreign exchange restrictions. It is common experience that even though adequate spare parts were provided while purchasing the equipment, sometimes breakdown occurs of some part which normally lasts a long time and hence was not carried in store. The situation partly arises from a lack of expertise in handling sophisticated equipment.

Lack of Infrastructure

Mineral deposits are often located in isolated areas. This makes it necessary not only to build housing, school, hospital, etc for the staff but also to extend railway and power lines. If this is not feasible, heavy duty road may have to be built and power generation facility established. In case the deposit is located in an arid region large quantities of water may have to be located, developed and brought from long distances by pipeline. No mineral project can bear all the above costs.

Role of Government

The extensive dependence of the industry on government in most developing countries makes it imperative that problems of this industry are clearly and dispassionately explained to various government departments and ministries. This becomes rather difficult because separate briefing is needed for the Ministries of Natural Resources, Industries, Finance, Planning, Railways, Labour, Commerce and, of course, Provincial Governments. Minerals are usually provincial subject and the authority of the Provincial Government is confined to leasing mineral properties and levying of rents and royalties.

It can be seen that there is conflict of interest not only between the mining industry and the government in some matters but also between various ministries also. To illustrate, the Ministry of Labour is primarily responsible for safety of labour working in mines. It is not concerned with either the quantity or the cost of production. It, therefore, often recommends the introduction of unnecessary sophisticated and expensive safety measures. Similarly, the Ministry of Natural Resources, with a view to promoting mineral development and production, recommends incentives like reduction or abolition of import duty on mining equipment, export, rebate on minerals etc. but the Board of Revenue being interested in increasing the collection of revenues is reluctant to agree to such concessions. The same applies to railways who wish to earn more and more from the mineral industry. This makes many low priced minerals uncompetitive in international market. The Provincial Governments consider mining industry as a source of revenue and levy high rent and royalties which become a burden on the industry. In the case of exportable minerals this burden cannot be passed on to the consumer.

SOME SUGGESTIONS

The starting point of solving the problems of mineral industry is a liberal and realistic mining policy. Drolet (1977), while discussing Mining Legislation and the Role of Responsible Authorities, he stated that a good mining code should state clearly and precisely the rights and obligations of the private investor and government and embody safeguards for the State's interests, discoverers' rights of development and tenure and for a fair return on investment with an equitable division of the proceeds.

Application of laws is often a matter of interpretation and, as such, minimum discretionary powers should be given to those responsible for enforcing the laws governing mining industry. The matter has been discussed in detail by a number of authors and an excellent collection has been edited by Laming and Ajakaiye (1977). After these general remarks breif comments are made on the problems discussed earlier.

1. Geoscience education. Geology, mining enginering, mineral dressing and metallurgy are taught extensively in the developing countries. However, most often there is a a gap between theoretical and practical knowledge. Countries with developed mineral industry have the opportunity of giving on-the-job training to their geologists and engineers. However, in countries with little mining activity there is a lack of practical training which adversely effects the quality of technical men coming out of the institutions. Those sent abroad for training on return find few avenues for the application of skills acquired and usually spend their working lives in offices. A detailed plan for education in mineral engineering coordinated with development needs to be drawn up and implemented. The participation of LDC geoscientists and mineral technologists in international workshops, symposia as also visits to mines and plants should be given high priority.

2. Mineral deposit evaluation facilities. Requests for government approval of the projects for implementation have to be supported by detailed feasibility studies for which extensive laboratory facilities are needed. In the case of metallic mineral deposits, in addition to geological evaluation, metallurgical tests on pilot plant scale are needed. Usually such studies are awarded to foreign consultants. However, by establishing pilot plant facilities in geological surveys or some central laboratory large scale tests can be done and various alternatives examined to select the best route and combination of products. Apart from reducing the expenses on feasibility studies, the pilot plant can be effectively used to demonstrate to the planners that valuable minerals can in fact be recovered from low grade ores.

3. Project evaluation expertise. Mining projects are different in most respects from other industrial projects usually set up in populated areas like textile mills, fertilizer plants, etc. If evaluated like ordinary industrial ventures, mining projects may fail to come up to the expectation as far as financial return on investment is concerned. The far reaching consequences of developing a mine employing hundreds of people in an otherwise isolated area are not fully appreciated. The mineral processing industry provides nine jobs indirectly for every job in the mine-mineral processing plant. Transfer of technology in a wide range of disciplines, acquiring self-sufficiency in vital industrial raw materials and development of areas which have no other potential are some of the major benefits of this industry which do not appear in the cashflows. In the case of low grade deposits it is not sufficiently appreciated that in addition to the usual benefits of an industry an otherwise useless

raw material, is being converted to useful products and in many cases earning foreign exchange. In fact, if for exportable minerals/mineral products cashflows are made only of the foreign exchange part, every mineral project will qualify for government support for development.

4. Financial incentives. Developing countries with agricultural economy do not usually allocate enough funds required to carry out meaningful mineral exploration and development. This is partly due to many national projects competing for meagre resources and partly due to long gestation period of mineral projects which are located in isolated areas where the development may go unnoticed by a large segment of the population.

Long gestation period of mineral projects also means that projects initiated by one government may come into production when their political rivals are in power who may exploit this situation to their own advantage. Hence, reluctance of government to commit funds to projects that may take more than say three years to complete. A conscious decision is, therefore, needed by according high priority to the development of mineral resources. In evaluating mineral projects only financial benefits are taken into consideration. Such evaluation usually shows low rate of return since the heavy cost of infrastructure development is also charged to the mineral project. The objective of all development is improvement in the quality of life. In this regard mineral industry can make a lasting impact on the standard of living of population in isolated areas though pure financial return may be low. Uniform development of the country as a whole is the goal of all governments and areas with no potential for the development other than minerals should receive a fair allocation of development funds.

High interest rates are a great burden on mineral projects due to long gestation period. Long term loans in local currency as well as foreign exchange should be provided to mineral projects. Many countries offer project oriented finances at specially low interest rates. Since mineral development initiates a chain of subsidiary industries governments should acord priority to mineral projects for such long term loans.

5. Laws governing mineral industry. Since mineral industry has to deal with many ministries and government departments, a lot of time is wasted in useless filling of forms, filling of returns and visits of officers from these departments to the mines/plants and offices. There is need to simplify procedures and programmes.

6. Lack of Engineering industry. Even though mineral industry can survive in isolation, a massive back up from engineering industry can enormously increase its profitability. While establishing engineering/fabricating units the needs of mineral industry should be kept in view.

7. Lack of Infrastructure. Many mineral deposits are located in isolated areas. Mining industry cannot be expected to bear the cost of this development. Roads, rail communication, water, power, school, hospital and training centres should be the responsibility of Federal and Provincial governments. Also, like agriculture, mineral industry should be provided electric power at specially subsidised rates.

8. Import duty and Sales tax. In most developing countries mining equipment is taxed at specially low rate. In parts of Pakistan all mining machinery is exempt from import duty and sales tax. Similar incentive should be introduced in all developing countries.

9. Mineral based industries. It is a common experience that even those mineral based products are imported whose raw materials occur abundantly in the country. Cement, fertilizer, plaster boards, activated clays, ceramics, table salt, paints etc. are common examples. Incentives by way of exemption from duties and sales tax on plant and equipment, permission to carry forward losses for say 10 years and exemption from income tax are some of the basic incentives needed. Price protection may become necessary in case of some products in the beginning and should be granted as a matter of right to all mineral based industries.

10. Consultancy. It is customary for Less Developed Countries (LDCs) to employ foreign consultants not only for feasibility studies but also for evaluation of mineral deposits. This usually works out to the detriment of the country. The consultant who does not have an indepth knowledge of the socio-economic conditions of the country may not be able to suggest the most acceptable strategies for development. It is, therefore, essential that local consultants are always associated with foreign consultants.

11. Mineral Development Advisory Group. To encourage mineral development and iron out difficulties of the mineral industry a Standing Committee of Geoscientists, Planning and Finance Departments' representatives should be set up. The Committee will act as the clearing house for all mineral exploration and development schemes. Once a scheme is cleared by this Committee its approval by government should be automatic.

REFERENCES

1. Proceedings of Symposium No. 2 Earth Science Aid to Developing Countries, 24th International Geological Congress, August 24-25, 1972, Montreal, Canada. Published by Overseas Development Section, Mineral Development Sector, Department of Energy, Mines and Resources, Ottawa, Canada.

2. Drolet, J.P., 1977 in AGID Report No. 8 below.

3. Laming, D.J.C. and Ajakaiye, D.E., 1977 - New Directions in Mineral Development Policies, Association of Geoscientists for International Development Report No. 8.

HUMAN RESOURCES DEVELOPMENT

THE TRAINING OF VILLAGE-RURAL GEOLOGISTS IN DEVELOPING COUNTRIES

M.B. Katz
Department of Applied Geology
University of New South Wales
Kensington, Australia

ABSTRACT

A village geologist training scheme is presented as a grass roots educational alternative complementary to the training of professional geologists. Strategies for the exploitation of mineral resources in developing countries usually place most emphasis on high tech, large scale exploration projects and often ignore the role of village-based prospectors. Recent socio-economic assessments question the big project approach to development and new directions are now exploring development at the human/community scale. Mineral resource development on the human/community scale involves the educational awareness of the villagers full mineral resource potential. This awareness would not only include small scale 'cash crop' mining of gold, tin, gems, etc., but would also concentrate on 'food crop' harvesting of industrial and fertilizer minerals, salts, building and ceramic material, energy sources, reliable water resources and the mitigation of geological and environmental hazards. This village geologist training program would contribute to decentralization and further decision making and self reliance. Instead of the tickle down process of development it would encourage new systems that are community based.

INTRODUCTION

The 1970 UN report on resource development, with particular reference to developing countries, mentioned the role that herdsmen and nomadic tribes can play as potential prospectors (UN, 1970). Assistance in development of village-level training programs to set up local cottage industries for minerals such as tin, gemstones, gold, etc. was a special recommendation of a recent workshop on 'Geosciences for Development - The Australian Role' (Katz and Langevad, 1985). As the most important resources of developing countries are the human resources (e.g. Bonis, 1982), the contribution of these individuals to mineral resource development can be important particularly if they have some training and are exposed to the fundamentals of the geosciences. Strategies for geoscience education in developing countries usually place most emphasis on tertiary level institutional building and on strengthening both the graduate and post-graduate sectors (Shimazaki, 1983). The role of geoscience education at the secondary and even primary levels is also receiving some attention (Stevenson & Armstrong, 1972). However a practical, direct and grass roots approach to this problem of geoscience education would be to establish training courses in the rural-village areas. These training schemes can be made to be an important and vital complementary activity to the more formal training of geoscientists.

GEOSCIENCE EDUCATION AND TRAINING

Deliberations on geoscience education in developing countries often discussed in many international meetings have stressed the importance of strengthening universities and other specialists-level training centres to produce more professional geologists (Katz, 1972; Cooray, 1977; Shimazaki, 1983). There is a tendency to strive for, and support geoscience education to the highest, elitist levels (Ph.D), but little data is available to suggest that this direction is the right, or most useful and relevant path (Ajakaiye & Woakes, 1983). This approach has produced many capable scientists in developing countries and many national and international organizations have fostered this group by encouraging and supporting further elitist-type ventures, meetings and training courses. There is no doubt as to the value of these projects in terms of satisfying the needs of this small community, but does the impact of these projects filter or trickle down to the fundamental development problems of resource assessment. The costs involved in training and sustaining these geoscientific elitist groups are very high and the benefits are not necessarily commensurate with the costs. International organizations must continue their support of the professional geoscientific communities in developing countries, however, they must also take the initiative and spread their support to lower technical and subtechnical levels of activity, especially to the uneducated, rural villager (e.g. Moravcsik & Exell, 1978, Freemantle, 1983).

For example, villagers could play a very important role as 'barefoot' geologists or prospectors/artisanal miners in locating a wide range of potential resources, as the legendary, and still active, prospectors/miners have done for metallic minerals in Canada, U.S.A., Australia, Finland and elsewhere, and in Japan where there is a long history of non-metallic prospecting/mining (Minato, 1983). Individual prospectors discovered the copper deposits of southern Peru, iron ore in the Nimba mountains of West Africa and diamonds in Tanzania (UN, 1970). These groups of prospectors can be encouraged, supported and trained at a very low cost and with potentially very high returns. Training courses for prospectors could run at a fraction of the costs of training professionals. In contrast, untrained prospectors and miners who are lured to "gold rushes" in many developing countries, the most remarkable group being the "Garimpeiros" of Brasil, lack geoscientific knowledge. They have difficulty in interpreting geological maps and reports and in their continual search for precious, get-rich-quick, deposits crisscross areas containing other valuable resources which they are not trained to recognize and evaluate. If they had some geological knowledge they would be in a better position to identify a wide range of resources in many inaccessible areas that would be an invaluable contribution to the country's resource assessment.

The villager-herdsmen have an intimate, instinctive knowledge of their environment and they are aware of the topography, water holes and rock types. These rural people often set up cottage industries in quarrying, ornamental and gem stones, rock crafts and building stones, and they are keenly aware of rocks of unusual colour, lustre, density, and breakage and shaping properties. Thus they can, with little training, be converted into village geologists trained in prospecting for economic and industrial minerals. In certain situations they also can be trained to take routine measurements on, and possibly maintain various geological, geophysical and geochemical equipment.

THE VILLAGE PROSPECTOR

Recent meetings and papers on resource exploration and small scale mining in developing countries often refer to the potential role that the uneducated villager can play as prospectors (UN, 1972; Meyer & Carman, 1980; Neilson, 1982; Salas, 1983; Wels, 1983). Wynn (1980) makes a strong case that one of the most important or critical responsibilities of a geological survey is prospector education. According to Wynn, two myths must be overcome. The first myth is that the prospector is an uneducated primitive. In reality these individuals can be highly motivated and technically capable. The second myth is that a local prospector can be taught exploration techniques but not to interpret the results. In fact, according to Wynn, he can be taught almost anything and is usually anxious to learn. These prospectors need only to learn to interpret data from the particular environment in which they are working. Fundamental to this training is the publication arm of the geological survey department of mines and a remarkable dispersion of prospector mining information is available (e.g. Wynn, 1980; Woakes, 1983).

In developing countries this need is not usually met, however, Rwanda, for instance, has published a book in the national language on mining exploration by artisans (Wels, 1983). The geological surveys in developing countries, with few exceptions (e.g. Legg, 1982), do not always have prospector education programs. As Ramachandra (1980) observes, in the case of India, neither the Geological Survey of India nor other government agencies offer any incentives to prospectors, although the survey have launched programs of making laymen mineral conscious by a series of demonstrations and lectures in remote parts of the country. Proposed courses for prospectors in Kenya are being developed by the Earth Science Technical Education Project of the Kenya Polytechnic (Martin, 1982).

Thus, potentially, one of the leading agencies for these training programs would be the geological survey. During surveys in various parts of the country the local villagers, herdsmen and even tribesmen would be encouraged to participate in survey activities and at the same time be trained in elementary prospecting/mining techniques. Ideally, when the survey completes its work it should leave behind trained cadres of prospectors that could continue their prospecting with promises of substantial rewards if they find anything. Tamale-Ssali a pioneer Ugandan geologist on exploration strategy proposed that the general public be involved in a mineral exploration militia (quoted in Mazimhaka, 1982). The geological survey could accomodate this program without any particular increase in general cost or time, and although the program would be scattered and sporadic it would be concentrated in mineral potential areas. Gaps in this program could be filled by University field research groups where staff and students investigations in remote areas would also have a compulsory element of training local prospectors. An example of this type of cooperation is given by Richardson (1982) in regard to a small scale mining project in the Mkomazi village, Tanzania.

The geological societies could also play an important role here by publicizing these programs of training and educating the rural populations to be aware of the mineral resource potential of their areas. Obviously state mining corporations and even the multinational companies should be encouraged to support these training programs. These prospector training schemes can be made to be an important and vital complementary activity to the more sophisticated and modern techniques of resource exploration carried out by these companies. In some countries regional centres of prospecting training or local schools of mines, possibly modelled on the Haileybury School of Mines, Canada (Carman, 1983) could be established to formalize the training on a more technical level (see Martin, 1982) and these schools could be located in well known mineral centres (e.g. Recuay Mining School in Peru, Wells, 1983).

Brock (in Berger, 1975, p. 25) describes a Mineral Prospectors Training program in Uganda with the cooperation of the Uganda Geological Survey, and these low cost, practical, non-prestigeous projects should be the kind of things that should be supported (see also Rao and Barros in Cooray, 1977, p. 40). The Finish experience of using amateur prospectors could be very useful in developing countries (Stigzielius, 1980). All of this of course is not new, and the Chinese have been involved in this sort of activity over the last several years. The self-reliant approach of the Chinese in their science and technology, with the aid of people on all levels, has developed the well known 'barefoot seismologist', trained in monitoring and predicting earthquakes (Davies, 1975), and this could be furthered in the search for mineral resources.

THE VILLAGE GEOSCIENTIST

Although there is still a nedd for prospecting for valuable commodities in developing countries, the present day metal glut and low prices places commodity-dependent developing countries (e.g. Zambia -copper) in an economic crisis. The developing countries resource needs have changed in the last few years and a more broad resource based background is required on all levels of national development and conservation (Cooray, 1984). This changing role of geosciences in development must be taken into account in the training of the village geologist. This includes the search for industrial and fertilizer minerals, salts, building and ceramic materials, alternative energy sources, clean and reliable water resources and the mitigation of geological and environmental hazards. In arid lands there is the major problem of desertification that also requires urgent geoscientific input to educate the masses on allaying erosion and land use mismanagement. One such changing role of geosciences in development is the important and relevant application of geology to soil improvement using locally available rock material (agrogeology) (Chesworth et al, 1983, Chesworth et al, 1985). The highly leached soils of the tropics are often deficient in minor nutrients and trace elements and have not been naturally rejuvenated by volcanism, sedimentation or other geological processes that add fresh rock to these highly weathered soils that increases the fertility of the land. In much of Africa, South America and parts of Asia where these natural processes have not been active on a large scale, a possible strategy for maintaining the fertility of the land would be to use common and readily available rocks as fertilizers.

The trends in developing countries for urbanization and the inherent problems that eventuate may now require geoscientific education among the poor masses so they will be able to understand the urban geological environment, especially in regard to water management and pollution. All these factors are raised to make the geoscientific educational program as complete, relevant and up to date as possible. The tasks are enormous, but a start in this direction, which could have some immeidate and tangible results, would be a village geologist training program for an appropriate, small scale, community-based, cottage industry level of involvement, that could, in the first instance, concentrate on local industrial minerals and rocks (e.g. phosphate for fertilizer), building materials (e.g. sand, gravel and adobe or mud brick) and energy resources (e.g. peat and low grade coal deposits) (Smith, 1982).

Village Geosciences in Sri Lanka : an example

Geology and geological processes influence the lives of all rural and urban communities in Sri Lanka in such important constructive areas as agriculture, water supply and industrial material and in potentially destructive areas as erosion, pollution and land slips (Cooray, 1984). Villagers utilizing local mineral, energy and water resources are a common sight, even to the casual visitor. Quarries are labour intensive and stones are sized for building, facing and paving down to road metal and aggregate. Lime kilns are common in the central areas where persistent formations of marble are located. Processes such as erosion, slope instability and land slips are often observed while travelling along the roads in the Sri Lankan highlands or when visiting tea plantations and hill stations. Sojourns to the beaches highlight problems related to coastal erosion, and stagnant, silted and polluted waters can be found both in the rural and urban areas.

The western half of Sri Lanka is located in a dry zone, an area of low rainfall and long periods of drought where problems of water supply are of major importance. The use of geology in siting good wells is vital for the communities in these dry areas. Unstable slopes in many areas of Sri Lanka lead to destructive and costly landslides and earth slips. Man made deforestation, inadequate drainage systems and cultivation of unstable slopes contribute to these disasters. Sri Lanka is largely an agricultural economy dependent on its soils. Geology controls the development and rejuvenation of soils and minerals and rock material may be used for fertilizer and soil improvement. Cottage industries in the use of clay for ceramics, stone quarrying and lime kilns are widespread and there are many local small scale gem, graphite and mica mines. The obvious mineral, energy and water resources are usually put into traditional local use, often with indigenous technology. As such the geological environment is not a complete unknown, but in most cases the use of geology for community and rural development in Sri Lanka is not maximised and many readily available resources are not utilized and potential environment hazards often ignored.

VILLAGE GEOSCIENTIFIC CURRICULUM

Large numbers of village geologists could be trained at little expense if an overall plan is developed by the educational/geoscientific/mining organizations of these developing countries. The role of these organisations would be to set up training courses in villages and to plan and coordinate all field training activities. The syllabus and detailed curriculum would be formulated with the aid of the educational planning authorities with special educational needs designed for the speedy training of semi-literate-illiterate rural folk. A possible, educational model to be considered is the Prospectors School run by the Saskatchewan Geological Survey, Canada, to provide native Indian residents of northern Saskatchewan with enough basic geological knowledge, so that they could work on mineral exploration field crews (Denner, 1976). This course is quite comprehensive and may not be suitable for developing countries, however an example of a simple, appropriate, draft syllabus is given in Table 1 (after Katz, 1983). Audio visual methods would be most important and special sets of minerals, rocks, charts, maps and other geoscientific educational tools and materials could be collected and assembled into portable kits. Recipe-type manuals and posters (e.g. Figs. 1, 2, 3, 4) using material, language and drawings which are clear to the public and appropriate for the content are being developed by the Global Learning Division of UNU/ICSU (United Nations University/International Council of Scientific Unions) (Friedman, 1984). These kits and manuals would be distributed to the geological surveys, university field research teams, state and multinational corporations and educational Institutes in rural towns and villages. The proposed AGID (Association of Geoscientists for International Development) Geoscience Video Education Program (GVEP) would be an important step in this direction.

GEOSCIENCE FOR DEVELOPMENT

The training of the village geologist would include the recognition of the enomalous behaviour of animals and plants. It is well known that animals are more sensitive than man to their geo-environment (e.g. Brooks, 1983).The story is told of the farmer in the west of Ireland whose ducks died mysteriously after feeding over parts of the property. The cause of death (lead poisoning) became clear when geochemical surveys leading to the discovery of the Tynagh lead-zinc mine, established the presence of a lead anomaly (Harkin, 1976). The Minister of Mines, Kenya, in his address to the Geological Society of Africa in December, 1982, at the Nairobi Meeting, also told the story of poor villagers that would have, as their favourite resting place, a large rock. Every evening they would sit on this rock and enjoy the cool evening breezes. After they had died in great poverty the rock was, by chance, turned over to reveal a rich gold seam. All these years they had been resting on a treasure and only if they had known to turn the rock over their whole life may have changed for the better.

In Third World countries the most important resources are human resources. The rural; village populations should be encouraged and trained to monitor their own environment for potential resources, as well as other important environmental matters, such as earthquakes, floods and landslides. If this village movement becomes operative (problems elucidated by Prakash & Rastogi, 1986), it would contribute to decentralization and development on a human scale and, ideally, would further the ecological balance, decision making and self-reliance on the small scale, local level. Instead of the 'trickle down' process of development it would further the aims of those few educational pathfinders that advocate new systems and counter movements that are community based (Wise, 1983).

CONCLUSIONS

These cadres, with limited geological training to supplement their inherent awareness of the land they inhabit would be a great and important source of general geological information. Remote and unknown outcrops would be identified and access routes established, and this would assist regional mapping and inventory surveys. Sites of instability, unusual phenomena, anomalous water levels and other harbingers of possible natural, and environmental disasters could be identified by these cadres and brought to the attention of the authorities. The geoscientific education of the rural folk would not only be a great contribution to locating mineral, water and energy resources in order to assess their own resources potential but would also contribute to the mitigation of natural geological hazards, increase of soil fertility and the improvement of the environment. With the assistance of UN agencies, non-government organizations such as ICSU and AGID and other interested groups like ITDG (Intermediate Technology Development Group) this program would further self reliance and the social and economic development of these countries. The current agricultural resource development problems of raising cash crops versus food crops can also be illustrated here in terms of analogous geological resource development. Geological 'food' crops fulfill the basic needs of the community, while geological 'cash' crops can be used by the community to augment their income. 'Food' crops consist of essential fertilizer minerals and rocks, building material and portable water. 'Cash' crops can be harvested in quarries and small scale mines and could consist of economic and precious minerals and building stone. This alternative development direction should not be at the expense of appropriate advanced technological progress involving more sophisticated mineral exploration and other geological-based projects (Harkin, 1976, Harada, 1978) and should be considered as a practical, fundamental, complementary activity.

REFERENCES

Ajakaiye, D.C. and Woakes, M. Training of Applied Geoscientists in Africa, Proc. Reg. Workshop. Role of Geoscience Educational Institutes in National Resources Development in Africa, CIFEG Occas. Publ 1983/1, 10-14 (1983)

Berger, A.R. (Ed). Geoscientists and the Third World. Geol. Surv. Canada paper 74-57 (1975)

Bonis, S. Problems and Requirements for Identifying, Exploring and Developing Resources, in Whitmore, F.C. Jr. and Williams, M.E. (Eds) Resource for the 21st Century, U.S.G.S. Prof. Paper 1193, 33-37 (1982).

Brooks R.R. Geozoology in Mineral Exploration, Episodes, 1983, 3, 27-2 (1983).

Carman, J.S. Training with Particular Reference to the Needs of Small Mines, in Woakes, M. and Carman, J.S. (Eds), AGID Guide to Mineral Resources Development, AGID Rept. 10, Bangkok, 47-55 (1983).

Chesworth, W. Marias-Vazquez, D. and Thompson, E. Agricultural alchemy : Stones into Bread, Episodes, 1933, 1, 3-7 (1983).

Chesworth, W. Van Straaten P. Semoka M.R. and Mchihiyo, E.P. Agrogeology in Tanzania, Episodes, 8, 4, 257-258 (1985).

Cooray, P.G. (Ed). Geoscience Education in Developing Countries, AGID Rept. 5, Bangkok (1977).

Cooray, P.G. The Geosciences and Resource Development in Sri Lanka, Pangea, 3, 38-41 (1984)

Davis, D. Earthquake Prediction in China, Nature 258, 286-287 (1975).

Denner, M.J. Prospectors School Manual, 6th Edition, Dept. Min. Res., Saskatchewan Geol. Surv., Regina (1976).

Freemantle, M. The Poor World needs Chemists, New Scientist, 28, 226-229 (1983).

Friedman, Y. 'Don't make new deserts ...' UNU Newsletter, 8, 1, 4-5 (1984).

Harada, H. Science: A Means of Locating Resources, Nature 275, 265-266 (1978).

Harkin, D.A. Systematic Mineral Exploration, Natural Res. Forum, 1, 29-37 (1976).

Katz, M.B. Relevance in University Earth Science Aid to Developing Countries, 24th Int. Geol. Congr. Montreal, Symp. 2, 182-186 (1972).

Katz, M.B. The Training of Village Prospectors for Natural Resource Development in Africa, Proc. Reg. Workshop, Role of Geoscience Educational Institutes in National Resources Development in Africa, CIFEG Occas. Publ. 1983/1, 53-55 (1983).

Katz, M.B. and Langevad, E.J. (Eds). Proceedings of the AGID-ILP Workshop, Geosciences for Development, The Australian Role. AGID Rept. 11, Bangkok (1985).

Legg, C.A. The Role of the Zambian Geological Survey in Small Mines Development, in Neilson, J.M. (Ed). Strategies for Small Scale Mining and Mineral Industries, AGID Rept. 8, Bangkok, 70-77 (1982),

Martin, W.J. Education of Middle level Earth Science Manpower in East Africa, in Neilson, J.M.(Ed). Strategies for Small Scale Mining and Mineral Industries, AGID Rept. 8, Bangkok, 171-176 (1982).

Mazimhaka, P.K. The Role of Small Deposits in Uganda's Mineral Production, in Neilson, J.M. (Ed). Strategies for Small Scale Mining and Mineral Industries, AGID Rept. 8, Bangkok, 61-70 (1982).

Meyer, R.F. and Carman, J.S. (Eds). The Future of Small Scale Mining, McGraw Hill, UNITAR, New York (1980).

Minato, H. Some Problems in the Development of Small-Scale Mines - The Case of Non-Metallic Mining, Role of Geosciences in Development, Proc. UNU-Tokyo Geog. Soc. Symp. 1981, 89-94 (1983).

Moravcsik, M. and Exell, H.B. Third World needs 'Barefoot' Science, Nature 276, 315-316 (1978).

Neilson, J.M. (Ed). Strategies for Small Scale Mining and Mineral Industries, AGID Rept. 8, Bangkok (1982).

Prakash, O. and Rastogi, P.N. Development of the Rural jPoor, Ifda dossier, 51, 3-16 (1986).

Ramachandra, A.V. Future of Small Scale Mining in India, in Meyer, R.F. and Carman, J.S. (Eds). The Future of Small Scale Mining, McGraw Hill, UNITAR, New York, 270 (1980).

Richardson, D.S. Small Scale Mining of Gypsum at Mkomazi, Tanzania, in Neilson, J.M. (Ed). Strategies for Small Scale Mining and Mineral Industries, AGID Rept. 8, Bangkok, 112-120 (1982).

Salas, G.P. Mineral Exploration and Exploitation (Small Scale Mining), Role of Geosciences in Development, Proc. UNU-Tokyo Geogr. Soc. Symp. 1981, 85-88 (1983).

Shimazaki, Y. Problems related to the Training of Professional Geoscientists with particular reference to Mining Resources Evaluation, Role of Geosciences in Development, Proc. UNU-Tokyo Geogr. Soc. Symp. 1981, 135-139 (1983).

Smith, R.G. Small Scale Production of Gypsum Plaster for Building in the Cape Verde Islands. Approp. Tech., 8, 4-6 (1982).

Stevenson, L.S. and Armstrong, J.E. (Eds). Geological Education, 24th Int. Geol. Congr., Montreal, Sect. 17, (1972).

Stigzielius, H. Small Scale Mining in Finland in Meyer, R.F. and Carman, J.S. (Eds). The Future of Small Scale Mining, McGraw Hill, UNITAR, New York, 249-251 (1980).

UN Mineral Resources Development with particular Reference to the Developing Countries. Dept. Econ. & Social Aff., New York (1970).

UN Small-Scale Mining in the Developing Countries. Dept. Econ. & Social Aff., New York (1972).

Wels, T.A. Small Scale Mining - The Forgotten Partner, Trans. Inst. Min. Metall. Sect. A, 19-27 (1983).

Wise, M.J. Education and Development, Role of Geosciences in Development, Proc. UNU-Tokyo Geogr. Soc. Symp. 1981, 128-134 (1983).

Woakes, M. Mineral Prospecting in Woakes, M. and Carman, J.S. (Eds) AGID Guide to Mineral Resources Development, AGID Rept. 10, Bangkok, 57-73 (1983).

Wynn, J.C. Exploration Geophysical Methods and Strategies and their use to Locate and Assess Small Mineral Deposits, in Meyer, R.F. and Carman, J.S. The Future of Small Scale Mining, McGraw Hill, UNITAR, New York, 75-81 (1980).

TABLE 1

VILLAGE PROSPECTOR TRAINING PROGRAM SYLLABUS

DRAFT SYLLABUS

CONTENT

1) Elements of mineralogy and lithology
2) Weathering, soils and fertilizers - agrogeology
3) Geological maps
4) Ore deposits and industrial minerals
5) Energy resources
6) Water resources
7) Construction, building material and engineering geology
8) Prospecting techniques - geological, geochemical, geophysical, geobotanical-zoological - surveying.
9) Geomorphology and the environment
10) Geological hazards

EQUIPMENT

1) Topographic maps
2) Geological maps
3) Airphotos - Remote sensing images
4) Stereoscopes
5) Traversing and survey equipment - pace, chain and compass
6) Handlens
7) Simple geochemical, geophysical, engineering and drilling equipment

SKILLS

1) Simple tests to recognize common minerals and rocks
2) Properties of economic and industrial minerals
3) Drawing and interpretation of simple geological maps
4) Use of compass, chain or pace for traversing - surveying
5) Operation and maintenance of simple geochemical, geophysical, engineering and drilling equipment.
6) Sampling techniques
7) Recognition of signs of geological hazards and environmental impact.

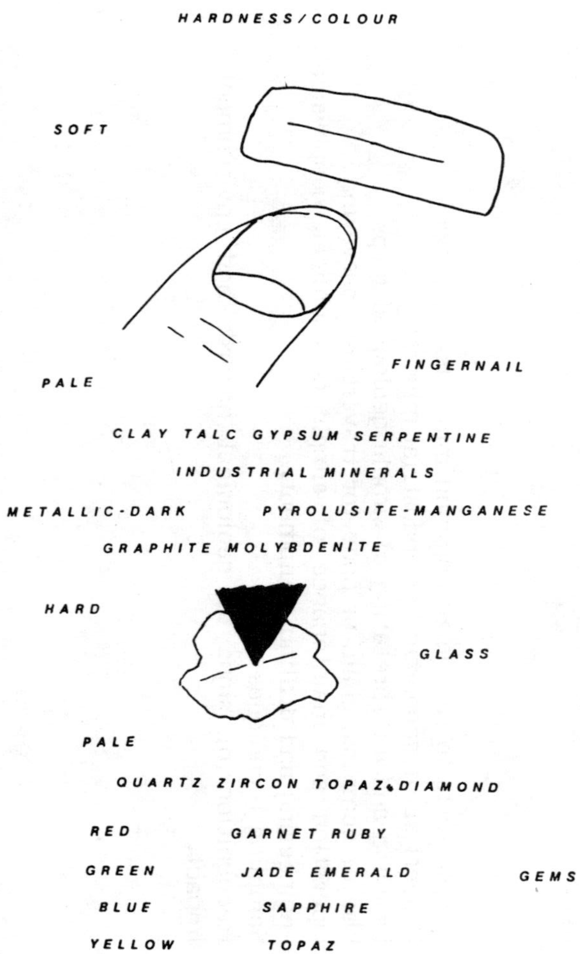

Fig. 1. Poster illustrating Hardness/Colour of Important Minerals

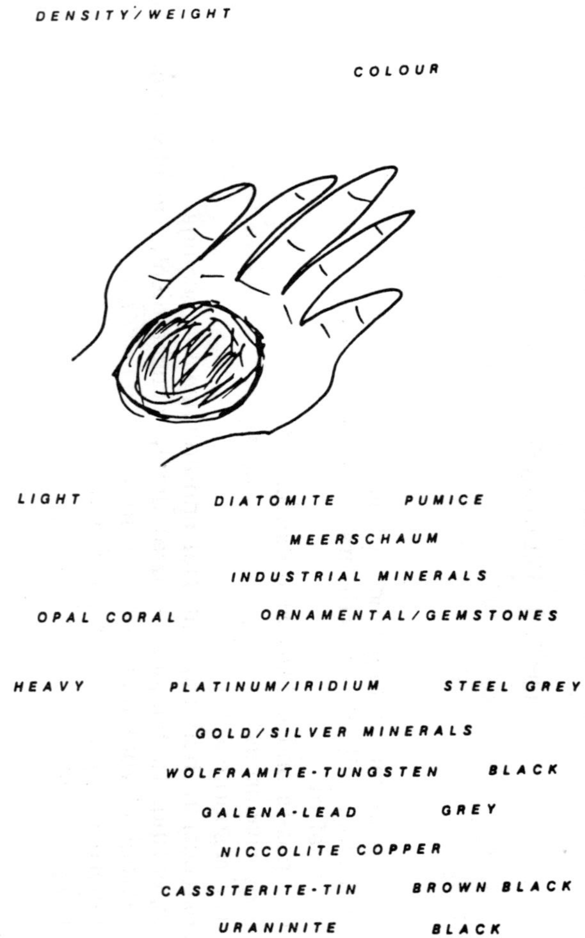

Fig. 2. Poster illustrating Density, Weight, Colour of Important Minerals

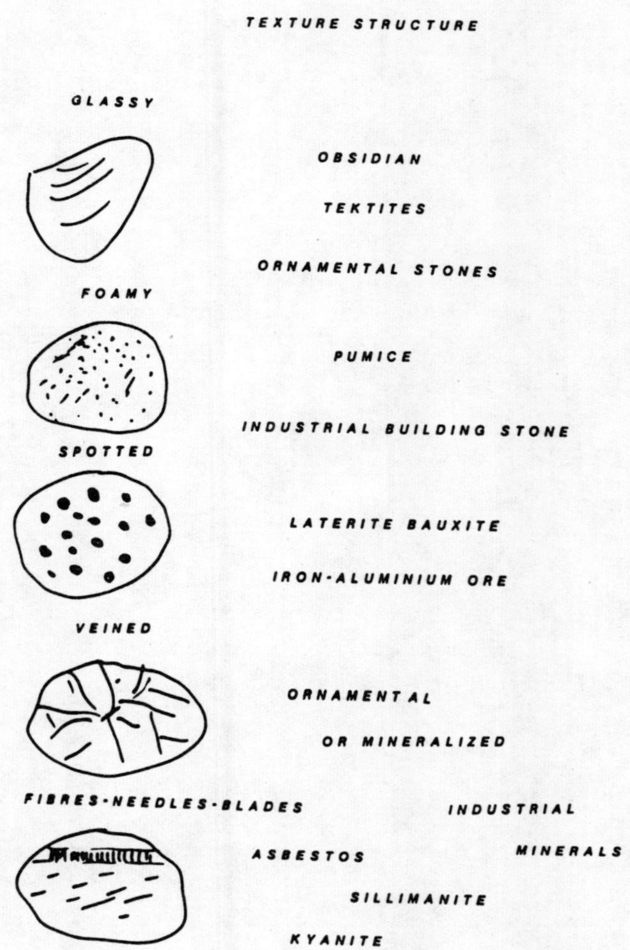

Fig. 4. Poster illustrating Texture/Structure of Important Minerals

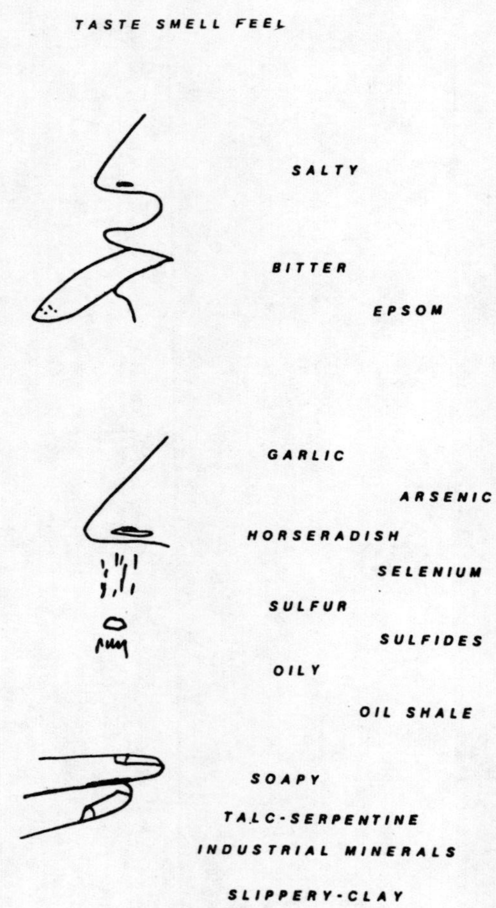

Fig. 3. Poster illustrating Taste/Smell/Feel of Important Minerals

MODEL OF GEO-DIFFUSION FOR EXPLOITATION IN LATIN AMERICA.

A Bhaskara Rao
Dept de Geociencias,
Universidade de Brasilia, 70910
Brasilia. Brazil.

Claudio de Castro
Dept. de Engenharia de Minas e CECINE,
Universidade Federal de Pernambuco,
50.000 Recife. Brazil.

Anand Rao Adusumilli
Social Communication Specialist,
SON 210,B1.A,Apt.423,
Brasilia.Brazil.

INTRODUCTION

With the advent of technological development, large quantities of mineral resources are being exploited for consumption. Thus, the Third World mineral resources potential continues to attract the attention of investors and mineral explorers from the First World. This has, at least in part, been responsible for the continued disinterest of the locals in the Third World to understand the importance of and necessity to explore and exploit the mineral resources to avail them locally.

MODEL OF GEO-DIFFUSION - THE PHILOSOPHY

A Model of Geo-Diffusion for Exploitation in Latin America envisages to waken the spirit of development of the local natural resources, through conscious approach of the people with confidence. The human resources are capable of creativity, labour, and productivity. The proper living habitat, if protected by the retention of the locals to the locality, both the space and social problems will be lessened and later tend to be eradicated.

When Man can occupy <u>well</u> the available space in his surroundings, then the Society will flourish and the Nations develop within their domains. This is the challenge of today's living.

EVOLUTION OF A SYSTEM

The basic point is to avoid desinterest and stimulate interest through the creation of consciousness to environment and riches that surround us. This is necessary and possible, but should be inducted from the very beginning of aquisition of knowledge and trials to live.

Training campaigns at the First and Second Grade in Schooling or even for the free-lance prospectors (garimpeiros) is an important starting point. Such an activity would require the following:

1) Diffusion of Geological knowledge or Geo-Diffusion, where Geo know-how and knowledge is made easily accessible. The word Geology seems to be not so attractive because, often, there is resistance for its incorporation in school curriculae. Thus it is preferable to use 'mineral

resources' or 'natural resourcess' as an attractive and markettable term. In fact, the geological community should learn how to market ideas they have.

2) Consciousness of the importance of locally available mineral resources, and methods how to explore and exploit them.

Several cases of small occurrences of clays, quartzites or pegmatites are simply ignored, and not availed of due to lack of such knowledge.

3) Concepts of local and regional development, with minerals or rocks as primary alternatives.

In arid regions where agriculture is often not so well developed, and cattle breeding has always problems, small scale mining is often the best option and alternative source of activity and economy.

4) Concept of joint community development and economies.

Small occurrences in a locality usually promote isolated mining groups resulting in loss of material (mineral) due to primitive recovery techniques. This is keasily avoidable if the groups organise a small community beneficiation plants with a sort of distribution of activities among themselves. Such a cooperative effort, when well conceived and practiced, will contribute effectively to the better production and participation in development and economy of the region.

5) Training of human resources, school teachers to begin with, to conduct this induction of consciousness, and to stimulate independent and local growth systems with the involvement of mineral resources. (See details in Training Campaigns).

6) Encourage and create alternative indigeneous technology as an option to retain the labour force to its own habitat with activity and production.

Some local state governments have started the Mineral Artesan Activity as Inductor Element in Micromining', through their Mineral Development Departments. In this case, some equipment for lapidation is also given, on solicitation, to the interested groups to support manufacture of stone jewellery.

Geo-Diffusion via Training Campaigns:

Educational technology has manifold styles, and appropriate educational technology is always necessary to formulate the optimisation of the outcome.

The basic step towards the creation of consciousness to explore and exploit mineral resources is through the training campaigns, with suitable operational system to make it part of a routine and easily accessible.

1) Material in-put:

Geo-Diffusion needs the practical knowledge of minerals, and rocks as their hosts or as utilitarian materials. This is possible through appropriate "kits" of minerals (30 in number) and common rocks (20 in number). Such kits consist of: a) small representative and frequent and economic minerals or rocks; b) a descriptive accompanying "guide book".

The samples should be representative in the sense that they are common in that country or of economic interest in general. If both the conditions are satisfied, then the collection becomes truly very representative for that country and beecomes a companion to the community participants.

The guide book should include a communicative and accessible description of the characters of the minerals or rocks in the kits, with small comparisions with similar material (minerals and rocks) that can be found. Guide minerals or typical rocks that characterise the presence of ore bodies could also be beneficial. The presence of the 'importance' and 'uses' of such minerals and rocks need to be given in a popular understandable language. With these qualifications and contents the guide book becomes an indispensable companion to the collection to make the kits very useful and instructive.

2) conceptual in-put

"Science and culture are not consumable articles, but they make us conscious of transforming articles to be consumable". The fact that Earth Science, through mineral and rock materials, is being induced to promote greater avail of locally available resources, makes the conceptual in-put as an important long-time effective tool.

Communicative, informative and scientific Conceptual Training Texts, specially prepared for this purpose will be the base of understanding the why, what and how of the training that is being imparted. Such texts need to be prepared viewing their importance to the mineral resources of each country, its dependency on mineral output of the nation, its participation in national industry, and its relevance in the locality.

This will avoid the dependency on the not easily obtainable foreign texts, which are not written for us, and which are either extensive or abstract for the objective of the Geo-Diffusion Model. In case of Latin America the problem gets crucial, but easy to overcome, because of the languages: French, English, besides Spanish and Portuguese. This is an essential part of the educational technology important for this Model.

3) Clientele:

The clientele for the Training Campaign is essentially Science Teachers from Schools, one from each selected School in one country, or region; in case of vast and populous countries like Brazil, Argentina and Venezuels. This is the initial training programme, and thus cannot be beyond 30 participants.

The training will permit them to be equipped with the 'kits' and with necessary knowledge to operate and 'diffuse' earth sciences to their pupils. They would constantly further their knowledge in Earth Sciences with the conceptual texts, already at their disposal. They would further receive 30 kits each with guides and texts to continue the same teaching and/or training in their regions or countries.

4) Duration of Training:

The duration of training period is full-time 6 days or 48 hours, with complementary night work, aided by kits with guide books and conceptual texts, besides the attention of the Teachers.

5) Feed-back:

With the growth in such consciousness, every local product, mineral or rock, becomes an object of curiosity and generates interest. The locals become a task force, thus, to look for obvious things and will try to use them with intelligence and creativity.

In place of ignoring what is available and near, the people start giving importance to the habitat, surroundings and will being. This should be the outcome in a short period.

CONCLUSION:

Propaganda, diffusion and marketing are tools of modern communicative system. Mineral resources are prime raw materials which need to be found and well utilised. They are the basis of industrial civilisation.

Within this framework, it is possible that several developing countries are on the search of ways and means to develop the local resources. However, if this method could have better and dynamic approach, from the grass roots, the outcome has to be far better.

This is the basis of the Geo-Diffusion Model. The authors feel, with their experience in different promotional activities, that this model is practicable. It can be started at a local, regional, country or continental level. With the knowledge of earth science stimulation activities in many countries in Latin America, they find that the progress achieved in such propagation is f ar from being commensurate with even basic necessities. It may be cited here that, even in the USA, the Geological Survey has always prepared folders trying to enthuse the citizens to look for minerals.

This proposal expects a Training Programme of five years to prepare well representative teachers in the Latin American countries. The follow-up will then be the country's responsibility. Thus, when once successful, it is proposed to enlarge its application in African and Asian countries, in case they donot intend to take a chance now.

It must be emphasised, finally, that such a project needs men with enthusiasm to prepare the kits with guide books and conceptual texts in an accessible and communicative manner to generate interest and pursuit. They are the backbones of the Model of Geo-Diffusion for Exploitation.

APPENDIX: I.

Some Supporting Information:

1) Mineral Collection kits, from several places, including the ones organised by: Instituto Nacional de Tecnologia, Rio de Janeiro(1957), Vlubr fr Mineralogia(1958),CECINE (1983-), at Recife, Brazil. They are even good for Geology Courses.

2) Propagation of popular Geology: Clube de Mineralogia, through Journal do Clube de Mineralogia, with chapters on Layman Mineralogy; and articles like: Autobiography of a Crystal; by author(ABR) since 1958.

3) Guide for the Museum of Minerals and Rocks,Recife (1966), with characters of minerals and rocks in very accessible language and inclusion of tables on the distribution of economic minerals and their abundance

in Northeast Brazil.

4) "Artesan Mineral Activity as Inductor Element in Micromining" Publication of Govert. of Pernambuco State (1985).

5) "Minerals of Brazil" 3 vols in Portuguese; with slide collection of 120, by Clube do Filme Fixo, Sao Paulo. 1974.

6) "Rock Collection" 20 samples, by CECINE (1984), Recife, as kits.

APPENDIX II.

SCHEME OF THE MODEL OF GEO-DIFFUSION.

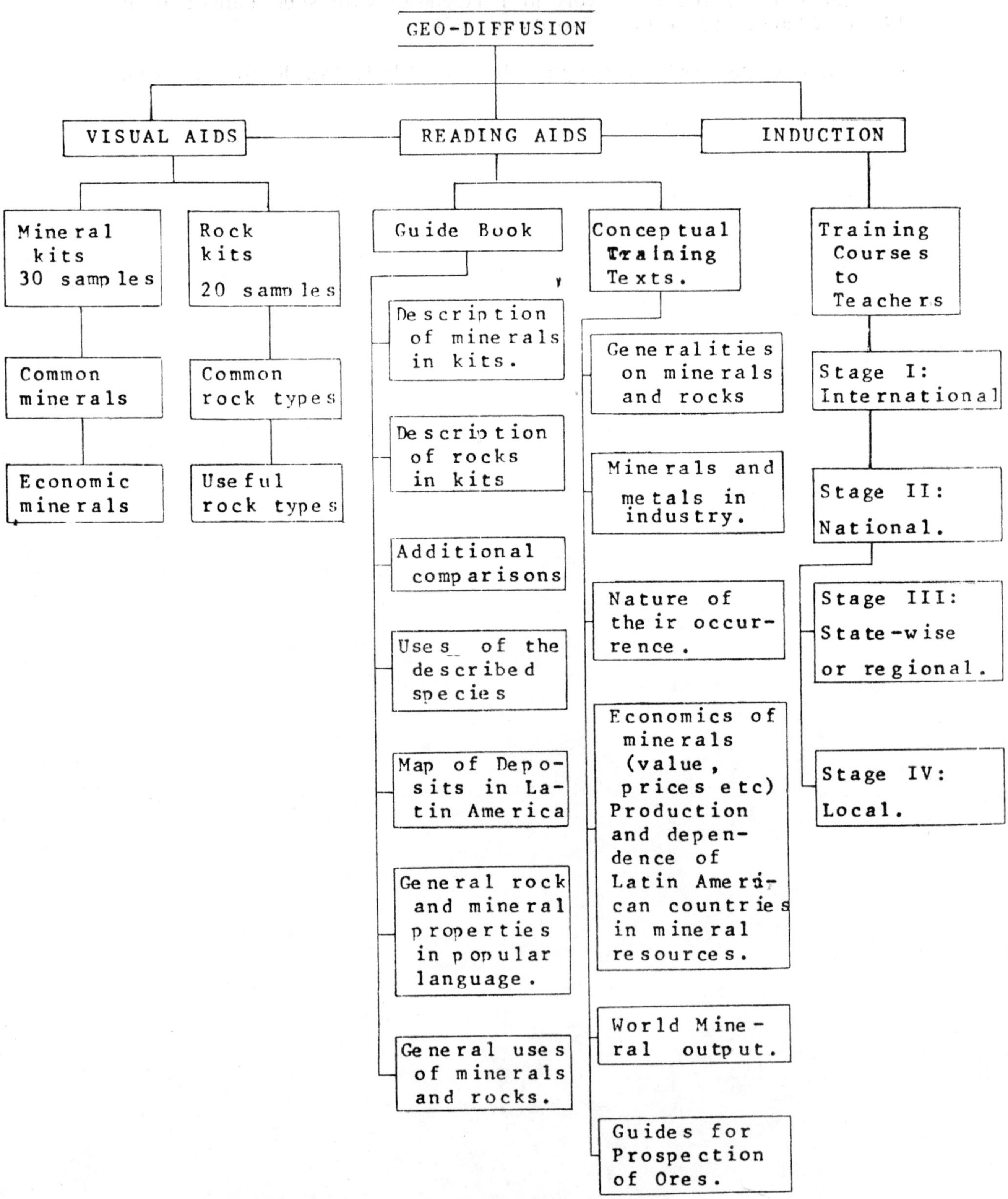

APPENDIX III.

TRAINING COURSES TO TEACHERS(INDUCTION CONTINUED).

Stage I: International Training Courses.

Venue: Latin American Countries, by rotation.

Duration: Five years.

Frequency: Twice a year.

Partcipants: Representatives ,atleast, one from each country, with a total partcipation of 30.

Expertise: International.

Sponsorship: International Organizations, and Govts. of Latin American countries.

Stage II:National Training Courses.

Venue: Each country's choice.

Duration: 10 years.,starting after two years of Stage I

Frequency: Twice a year.

Partcipants: At the discretion of the Govt. but with representation of one from each State or region, at least, and the maximum to be decided.

Expertise: National, with already two to four to guide the courses.

Sponsorship: Local Government and OAS.

Stage III:State Training Courses.

Venue: A town in the State.

Duration: Depending on the demand. for at leat 10 years, starting after one year of Statge II.

Partcipants: State school teachers.

Expertise: National.

Sponsorship: State Government.

Stage IV: Local Training Courses.

This stage is only found indispensable in vast countries like Brazil, Argentina and Venezuela.

This stage is similar to Stageg III, with interest of the National Foundations for the Development of Education

There should be a constant feed-back and follow-up, or otherwise, new teachers will not gain the same momentum.

EXPLORATION AND EXPLOITATION MENTALITY IN HUMAN RESOURCES TRAINING (MANPOWER TRAINING)

A Bhaskara Rao
Dept of Geosciences,
University of Brasilia
70910 Brasilia, BRAZIL.

INTRODUCTION:

Manpower is the only perennial treasure of any country. Any effort towards perfection in its training is only beneficial to the Nation. This is true for Developing Countries where the demand on expertise is continuously on the increase to attend to the multiple problems provoked by industrial projects and inadequate national economics. The demand on the raw materials for the industry could be attended to with contribution of the earth sciences. This is possible with a exploration mentality, and consciousness towards mineral exploitation as raw materials to sustain the industry.

EXISTING SYSTEM:

Geological and mining engineering manpower in developing and developed countries has a traditional system of education, based on the following essential aspects:

1) Theoretical systematics, which conduct the capacity of understanding the evolutionary aspects of the topics treated.

2) Practical verification, in the laboratory and some field work, to understand the facts known in theory.

3) Training in a general pattern to establish some applicability of the knowledge in " basic mapping" or "mine operation".

It may be said that these have been basically good enough to produce professionals in the earth sciences. It is such a basic mapping and mine operation that has been useful to promote the progress that occured in the developing countries.

On the other hand, the technological advances that are being created in developed countries are being adopted for applied work, such as exploration and exploitation in developing countries. The mineral rushes in Canada, Australia and Brazil have shown the potentialities of these countries, and have stimulated more exploratory work elsewhere.

NEW CENTERS AND COURSES:

Undoubtedly, this has given the impetus for the creation of new Centers and Courses in exploration. But, however, it must be said that the exploration consciousness in the developing countries is not adequately promoted during the manpower training. This is primarily attributable to:

1) The traditional approach of geology to understand the basic sciences and geological subjects like mineralogy, petrology, structural geology, stratigraphy, photogeology and economic geology.

2) The constant pursuit of these sciences by the staff and researchers,

and the normal dissociation of their applicability to other sciences or to locate mineral deposits.

3) The constant inflexibility in the objectivity in geological studies, viz. isolated geological works with detailing of stratigraphy or structure or mineralogy etc.

4) The lack of efforts to integrate and induce a wider spectrum of approach towards stimulating "returns" for the "investments" in geological sciences, through pursuits or efforts for the discovery of ore concentrations or water resources or energy resources.

5) The continuous enthusiasm towards more philosophical than realistic dissertations, and non-conclusive tendency in geological scientific works.

EXPLORATION: TENDENCY AND CONSCIOUSNESS

Thus, it is often noted that few only generate a tendency for exploration work which needs:

1) Initial curiosity and doubt in every small occurrence, guide or evidence to confirm its basis and its importance.

2) Comprehensive understanding of the basic and applied geological sciences, their applicability and their importance in the trials to locate target areas.

3) Critical accumen to select methods and means for operational work to demarcate promising mineralised or rich zones of resources.

4) Capacity to take risk and to make decision to reject or proceed at every stage of operation, in exploratory campaign.

Since exploration is finance-based, as any other investigation, it should involve the least technical risk. It is possible, only, when a geologist is well prepared in other subjects specially in economic geology (ore deposits, water resources, or energy resources) which require good concept of structural geology, mineralogy, and petrography, besides stratigraphy, field geology and tectonics.

In this context, it is always the great complaint of professional enterprises, that manpower coming from the Universities is not well prepared for exploration work. This has promoted the creation of Technological Universities or Institutes of Technology with the result that some technical improvement has been imparted, but at a loss of a universal vision, basically induced for living, in the traditional Universities(1).

It is necessary to emphasise here that if manpower training is to attend to the development envisaged by the Third World countries, the Centers of Education and specially the teaching staff should stimulate a mentality of exploration and exploitation work, even from the very beginning of the geological education. Geo-diffusion at the earlier stages in the schools, with kits of minerals and rocks and guide books, will as well be a part of the system (2).

To achieve this, special efforts are not a pre-requisite, but involves the following -

1) Generate interest to the fact that "all investment expects returns". This includes earth sciences, and the returns in geology are the location of ore deposits, water and energy resources, which

are the products. Their exploitation assures new investments.

2) Call attention to the importance of minerals, rocks, water, energy resources, their utilities; the contents of metals in minerals; the abundance of resources in the country and world. This could be given in mineralogy, and economic geology; and detailed in hydrogeology, energy resources and mineral economics, if such subjects are taught.

3) Stimulate interest in even a small occurrence or presence of "economically interesting mineral" in any field campaign, and devote some attention to record its extension, participation, and behaviour. Such a procedure should become part of a routine observation, and integral part of the field work, with the result that discoveries could become more frequent.

Among the many examples, two may be narrated here as illustrations. i) In the Final Year Geological Report that involves 25 days geological mapping of an area, occurrences of chalcopyrite have been mentioned with no single additional data. The student was told by the Professor that their objective is geological mapping. (Later campaigns <u>discovered</u> an extensive bed with a participation of well over 1% of Cu in this occurrence, and feasibility studies are now on the way.) ii) One student making a petrography dissertation ignored small occurrences of W ore, both in field work and in the final report, as they are not the object of his studies. (A prospector with a Mineral Light-UV Lamp Short wave, <u>discovered</u> a very rich deposit in the area, and not even a reference was made to the dissertation.)

4) Involve "curiosity" to determine a mineral during a laboratory or field work, (Tantalite samples were ignored and thrown away in field as another hematite.) and make known normally the prices of some minerals.

5) Dissuade the use of "opaques" in field and research reports, and encourage a doubt when such opaques are in quantities (even 2%) in some rock types. If the opaque is chalcocite it signifies a 1.7% Cu in the rock. Some reports even mention the presence of 10% opaques, with no mineralogical determination (or even a trial, or test), attributed to magnetite.

6) Stimulate interest in defining rock types and not the **usage** of granite, schist and gneiss. Some schists (chlorite) are so important in an appropriate geological context that they could be guides for exploratory work. Rapakivi granite, once a petrological curiosity and complexity, is today a possible repository of Sn ores.

7) Stimulate the importance of the genesis of minerals. It is common to see that geologists spend hours discussing the genesis of ore deposits. If this could permit them to postulate a "genetic model" for the case under study, the contribution will be very useful.

A genetic model is not to be conceived only as a theory on the evolution of the deposit. It is the basis to visualise several aspects of interest, such as: nature of the body, nature of the mineralisation, structural behaviour of the mineralised zone, distribution styles of ore minerals, paragenesis and associated ore minerals and gangue, and even textural and structural patterns of the ore minerals.

Such a prognosis induces immediately the structure of an exploration programme.

8) Induce the formulation of exploration models, in each case, based on the genetic model.

It is common in geology to use the terms 'probably' and 'possibly', which donot involve exactness, since geology is a natural science. But in exploration and exploitation, both definitions and decisions are necessary at the end of each stage in operation.

Thus, an exploration model needs at each stage the decision-making as to proceed further with the investigation or change the system of operation. This may also involve, at a latter stage, the change to another genetic model. However, in the first two stages of exploration, the validity of the genetic model used for the formulation of exploration model, will be clarified. Since exploration work involves strategies and systematics of operation, as any other creative research, it is vulnerable to such small changes. This fact doesnot impede the postulation of an exploration model, but on the contrary, it reforces the necessity to have such a conceptual procedure.

Training manpower within this approach, permits the cultivation of creativity and selection of options.

Since exploration model is postulated in each case, it avoids repetetive procedures and operation of conventional systematics like: geochemical, geophysical, topographical, tr enching, pitting and drilling operations, in all types of exploration works. Some of such methods donot work in a specific genetic environment, and as such are excluded in the early stages. The result is less operational costs, less time and more objectivity.

It is frequently noted that a government sponsored exploration programmes is bulky in finances, duration, methods, and experts. On the contrary, a private enterprise has less time, less men, and uses few methods and less finances to get the best results towards the objectives.

9) Induce a constructive critical attire towards exploitation of resources. It is often noted that manpower from Centers of Training is not aware of the factors that involve a "feasibility" study of any deposit which has its reserves measured, inferred and indicated.

The developed countries have their systematics in making feasibility studies with the total participation of production economists and engineering economists. But in developing countries, even a small economic feasibility study becomes a bottleneck in several mineral development projects. The basic concepts discussed, often, will permit a consciousness of exploitation problematics of mineral resources. This is indispensable in manpower training.

A simple illustration of a case is of interest in this context. In a case of a supposed diamond occurrence in Amazon region, one student preferred the paving of a road of 1000 km to transport the kimberlite mass to the town and beneficiate diamonds. This is a typical case of lack of concept of beneficiation, and diamond transport by air.

CONCLUDING REMARKS

"Underdevelopment is not a status, its state of transition". The developing countries primarily need to induce the development consciousness and avail the talents to formulate the best use of indigeneous technology. Vast terrains are still unknown in many countries, and their potential riches are still targets of investigations.

Exploration and exploitation are the only means to produce the prime materials for the industrial expansion envisaged by all independent nations. If this has to be accelerated; the only solution is the training of manpower with these objectives. No one desires that all the geologists turn out exploration work. But if they have in mind the necessity to look for the hidden riches, then the outcome, possibly, would be better.

The induction of the mentality towards exploration and exploitation of mineral resources is needed in all countries, but imminently in the developing countries. It is a fact most developing countries still donot have a reasonably good geological coverage of their tarritories. In attaining this, they could also foster greater integration, absolutely indispensable, with exploration. This needs a little more effort and time, but with the mentality to turn science to technology - a process fundamental for development.

As mentioned above, exploration needs mentality, trained from the very beginning of the educational career, not only in the University, but also from the School period. If the value of minerals, their importance in the industry and for the economy of the country are propagated, then the entire process of development of earth sciences will be a normal systematic affair in any nation. However, in the University education the mentality should start with the teachers of all geological subjects. If they donot accelerate this dynamics, the developing countries will continue facing difficulties and will have to depend on foreign expertise for times to come.

In conclusion, it may be said that for developing countries and their modern industrial intentions, the growth in basic raw material production is indispensable. Geological exploration and exploitation is one of the basic ingridients.

REFERENCES:

1) Bhaskara Rao,A.,Egboka, B.C.E. & Onyekwe lu,L.,1983. Technological Universities, Natural Resources and National Development.,AGID NEWSS, Bangkok. No.34,p.1-3.

2) Bhaskara Rao,A.,& Castro, Claudio.,1986. Model of Geo-Diffusion for Exploration. (Proposal submitted to Unesco).

ENVIRONMENTAL ISSUES

ENVIRONMENTAL STRATEGIES FOR EXPLOITATION OF MINERAL RESOURCES IN DEVELOPING COUNTRIES

S P Banerjee
Department of Mining Engineering
Indian School of Mines, Dhanbad, India.

ABSTRACT

The developing countries are faced with the twin problem of improving the living standard of a growing population and at the same time preserving the environment. Intensive exploitation of natural resources, including mineral deposits helps in economic development but cause damage to the environment. In the developed countries, mine operators have to comply with stringent environmental protection laws. Developing countries are also gradually introducing such restrictive legislative measures. But the models in a developing country has to be different at least at the present time, as the minerals are being sold in a buyer's market and capital for development of mineral deposit has to be attracted. Therefore, the priorities of environmental protection measures should be worked out and only essential precautions against deforestation, land damage and water pollution may be enforced in the beginning. Education to bring in environmental awareness and self-regulation by the mine operators should be the backbone of such an environmental policy. The motto should be development with conservation and not conservation at the cost of development.

INTRODUCTION

Minerals are essential for sustaining the economy of both developed and developing countries. In case of the latter, however, the exploitation of mineral resources is of relatively greater importance. The developing countries are chronically short of foreign exchange of which minerals provide a valuable source. There is usually a high growth rate of population and an urgent need to raise the standard of living of the masses - both factors needing an accelerated economic growth and a progressively larger quantity of minerals. Exploitation of minerals also provide an important source of employment for the indigenous population and help in development of remote areas in the developing countries.

However, exploitation of minerals and the associated processing activities often cause great damage to the environment. The concern for protection of the environment has become widespread in recent years. The aim of this paper is to examine the environmental issues involved in exploitation of mineral resources, how priorities can be different in the developing and the developed countries while dealing with the environmental issues and to discuss the merits and demerits of the various strategies which can be adopted to contain the environmental problem.

POPULATION GROWTH & ENVIRONMENTAL DAMAGE

Lester Brown, in his State of the world - 1985 Report (Brown, 1985) has shown how the world of the mid-eighties is a world of **stark** demographic contrasts. Some three fourth of the current world population of 4800 million live in the developing countries. The population sizes of a dozen and half countries, all developed and in Europe, are now almost stationary. On the other hand, the population of the developing countries are growing at a fast rate varying from 2 percent to some 3.7 percent per year. In most developing countries the struggle to improve living conditions while keeping up with record rates of population growth has been a losing battle. There has been tremendous population-induced deterioration of forests and soils in the developing countries, specially in Africa. The leaders of these countries are faced with a difficult situation. The immediate need for increased food, energy and other resources is breaking down the centuries old stable relationship between people and environmental support systems. Many third world leaders are now speaking out on the urgency of slowing down population growth but this will need time. Most demographers project that the world population will continue growing until it reaches some 10 billon and the increase will take place mostly in the third world countries, many of which will experience several fold increases--in their population. Inevitably, there will be tremendous stress on the natural resources and the life support systems in these countries and maintaining the quality of environment will be low down in the list of National Priorities.

DIFFERING MINERAL RESOURCES & ENVIRONMENTAL PROBLEMS IN DIFFERENT COUNTRIES

The mineral resources of the world is not uniformly distributed. The concentration of the major oil deposits of the world in only a few regions is a prime example of this localised occurrence. What is true of oil is also true for most other minerals and gives rise to the interdependence of nations in the matter of meeting their needs of minerals and metals. The developing countries as a whole have a fair share of the world's mineral resources. One estimate puts it at 38 percent for the developing countries, when the Western market-economy countries posses 35 percent and the centrally-planned-economy countries have 27 percent of the resources (Butt, 1982). This share of the mineral resources in the developing countries is, however, small when considered in the background of the high proportion of the world population living in these countries. Again the picture is not uniform within the developing countries themselves. Within the LDC group there are countries such as India and Bangladesh which are densely populated and well explored, and also the countries in Africa and Latin America many of which are relatively sparsely populated and not fully explored as yet for their mineral potential. Some of these countries have extraordinarily rich mineral deposits (e.g. Chile, Zambia and Papua Newguinea for copper deposits), supporting a thriving mining industry which contributes greatly to the national economy. The environmental problems created by mining are also different for different deposits and in different countries.

ENVIRONMENTAL PROTECTION MEASURES - THE HISTORICAL DEVELOPMENT

The concern for protection of the environment from various anthropogenic activities became widespread in the developed countries in the 1950's and 1960's. The Opencast Coal Act (1958) of U.K. require restoration of mined site. The General Mining Act (1965) of Northrhine-Westphalia of FRG requires operators of brown coal mines to get a reclamation plan approved by the authorities. The National Environmental Policy Act (1969) of USA requires an EIS (environmental impact statement) to be prepared for all major federal actions significantly affecting the quality of human environment.

In the developing countries, movement for environmental protection effectively started after the UN Conference in environment held in Stockholm in 1972. The leaders of the Non Aligned Movement in the Delhi Summit in 1983 expressed the need for protection of the environment in the developing countries. India, as a leading Third World Country, has introduced legislations for prevention and control of water (1974) and air pollution (1980). It has also introduced a system of requiring an Environmental Managment Plan to be prepared before new mining projects are cleared by the Government.

DIFFERING PERSPECTIVE OF ENVIRONMENTAL PROBLEMS

The public pressure groups for environmental preservation has been active in the developed countries for many years. The last decade has seen the emergence of similar active voluntary groups for protection of the environment in the developing countries as well. The 'Chipco' movement to protect the trees in the Himalayan foothills is a glorious example of successful public participation at the grass root level for protection of the environment. The struggles to ban limestone mining in the Mussoorie hills of U.P. and bauxite mining in Gandhamardan hills of Orissa on environmentral considerations are continuing. Since 1972, the Government of India has taken many measures to protect the environment in the country including in mining areas, but the clamour for still greater environmental protection measures in mining areas are often voiced by the Public and the Press and cases of stalling of mining projects due to environmental considerations have occurred. The examples of the prevailing more stringent legislations in the developed countries are often cited in this context. For instance, reclamation of mined land to the original topography is a statutory requirement in many developed countries. While making similar legislations the policy makers in the developing countries must consider the differences in the existing situations between the developed and the developing countries. In the developing countries, the grazing pressure is extremely high, and efforts towards vegetative reclamation needs the extra cost of erecting and maintaining fences. Similarly the hot and dry climate in many developing countries make reclamation a more difficult and expensive task than in the temperate regions with regular rainfall throughout the year as found in most of the developed countries. The high per capita consumption of energy and other material resources and concentration of industries have contributed to a high base level of environmental pollution in the developed countries. Naturally, any further deterioration of the environment in these countries may cause danger to the population. The developing countries, on the other hand are faced with a different kind of environmental problem - lack of sanitation, accumulation of wastes, deforestation, dereliction of land - all directly stemming from over-population and lack of development. Mrs. Indira Gandhi (Gandhi, 1972) in her speech at the 1972 Human Environment Conference at Stockholm said

"The environmental problems of developing countries are not the side effects of excessive industrialisation but reflect the inadequacy of development. The rich countries may look upon development as the cause of environmental destruction, but to us it is one of the primary means of improving the environment for living, or providing food, water, sanitation and shelter, of making the deserts green and the mountains habitable".

Obviously there is a difference in the perspective of the two groups of countries.

THE WORLD RECESSION AND DECLINE OF THE MINING INDUSTRY IN THE 1980's

The world economy is going through a period of recession since the oil price hike in the seventies. At present there is a considerable over supply of many minerals including copper, iron ore, uranium and coal. the third world countries whose economy is significantly dependent on the export of minerals and metals would not like to price themselves out of the foreign markets particularly at this juncture. Only low-cost high priority environmental protection measures can be adopted at this stage leaving the costlier and less urgent measures to be implemented when the economic climate improves.

THE NEED OF A STABLE AND SIMPLE ENVIRONMENTAL PROTECTION POLICY

Over the years the principle of economy of scale has been applied in exploitation of mineral deposits and present day mines are huge capital intensive projects with long gestation periods. The developing countries often find difficulties in organising finance for such projects on their own and approach multi-national companies or leading institutions for financing the projects. Recurring threats of nationalisation or stiff environmental regulations may unnerve the financing organisations. A stable environmental policy with a clear perspective of the legislations to come and standards likely to be fixed can help in removing the misgivings in the minds of the investors.

The third world countries also suffer from another weakness - a generally inefficient administration. The proverbial red tape is much in existence and development projects often suffer from bureaucratic delays. The system of environmental control of mining projects through the issue of licenses and permits for the mine is a model often followed in the developed countries. If a similar system is introduced in a developing country, precautions must be taken to ensure that this does not increase the bureaucratic red tape in clearance of the projects. Similarly the LDCs should be cautious in creating new Government agencies to monitor the implementation of environment management plans in mines, as in the Third World such agencies often tend to thwart development efforts and become breeding ground for corruption. They may achieve the same goal by having some sort of self regulation scheme in the mining companies through education and increased awareness of the mine operators.

EMPHASIS ON SELECTED ENVIRONMENTAL PARAMETERS - PRECAUTIONS AGAINST DEFORESTATION, LAND DAMAGE AND WATER POLLUTION

For deciding the priorities of environmental protection measures to be adopted in the developing countries one must first list the various environmental damages caused by exploitation of the mineral deposits. These include deforestation, land damage, visual impact, pollution of surface water, disturbance of the surface and ground water regimes, air pollution, noise pollution, damage due to blast vibration and air blast, subsidence, disturbance to wild life and loss of rare species of fauna and flora. These have been discussed in detail elsewhere by Banerjee and Marwaha (Banerjee and Marwaha, 1984).

In the opinion of the author, land damage, deforestation and water pollution are the three most important adverse impacts of mining on the environment and these might be tackled first by the developing countries leaving other aspects such as visual impact, noise pollution, air pollution etc. to be tackled later unless the mine is situated near an urban centre. All the three aspects can be tackled at a reasonable cost if plans are drawn for the same at the initial stage of mining. Only the minimum area in a leasehold should be bared at any time, with further deforestation progressing only with the progress of the mine face. As far as practicable no tree should be felled in constructing the housing colony for the project. The mine planning should be so done that reclamation is carried out simultaneously with the mining operation. The backfilled land should be revegetated as early as possible and there should be no compulsion to grade the backfilled area to the original topography. Restoration to agricultural or pasture land or creation of water reservoirs or other alternative uses of the mined area should be decided at the planning stage.

The precautions against water pollution will vary with the mineral mined, the type of mining and the site conditions. The run-offs from surface mines usually have a high silt content and the water should be led to sedimentation ponds before discharge. In a mine with a mill or a washing plant, the water should be recycled. Where the problem is of acid mine drainage due to oxidation of pyrites, the toxic layers should be quickly buried to a sufficient depth and the water taken through a treatment plant containing aeration and neutralisation ponds. Where heavy metals, cyanides or other chemical reagents in the mine effluent cause problem to the downstream waterbodies, suitable water treatment facility has to be created. The quality and user pattern of the receiving water body should always be considered while deciding on the water treatment facility to be installed. In the developing countries, river and lake water is often directly used for drinking and other domestic purposes by the villagers living in down stream regions and concentration of pollutants in the mine effluent should be controlled.

CONCLUDING REMARKS

The environmental problems faced by the developing countries have many dimensions. Existing overpopulation, coupled with a rapid population growth and poor living standards are the root cause of the degradation of the environment. These countries have now embarked on varied degrees of industrialisation programmes including exploitation of the mineral resources. These development activities no doubt impinge on the quality of the environment. However, if the developmental activities are not pursued, the environment in these countries will still be subjected to increased stress and deterioration from a rising population living below the poverty line. If all the high cost environmental protection measures as adopted in the developed countries are enforced on the mining industry in the LDCs, the danger of pricing their minerals out of the international market may arise. Therefore, cost benefit ratio of the various environmental protection measures must be considered and priorities worked out. The objective should be to have development with conservation and not conservation at the cost of development.

REFERENCES

1. Brown, L.R. et al "State of the World 1985" Prentice Hall of India, New Delhi, 1985.

2. Butt, N.A. "Future Mining Supplies and the LDC" Trans IMM, Vol 91, Jan. 1982, PA11.

3. Gandhi, Indira "Man and his Environment" address at the Plenary Session of UN Conf. on Human Environment, Stockholm, Sweden, June 14, 1972.

4. Banerjee, S.P. and Marwaha, G.S. "Environmental considerations in exploitation of Mineral Resources with special reference to the conditions in Developing Countries". XII World Mining Congress, Paper 3.10, IE(I), New Delhi, November 1984.

ECOLOGICAL DISTURBANCES ASSOCIATED WITH EXPLOITATION OF LOW GRADE ORES IN LESS DEVELOPED COUNTRIES

B B Dhar
Department of Mining Engineering
Banaras Hindu University, Varanasi
India

INTRODUCTION

The history of human civilisation has shown that mineral resources have played and will continue to play an important role in the progress of mankind. The dependence of primitive societies upon mined products is illustrated by the nomenclature of the epochs like Stone age, Bronze age and Iron age, a sequence which also shares the increasing complexity of society's relationship with mining.

In the beginning, man extracted only a few minerals that were of relatively of high grade. However later, the use of scientific methods to develop mining and mineral processing techniques helped in exploiting low grade ores, both in developed and less develop countries. Keeping in mind the economics and demand, mining has to be carried out on a large scale. But at the same time low-grade ores and small mineral deposits cannot be neglected, and have to be exploited economically at least in less developed countries.

Mining of small mineral deposits, if exploited economically, would generate prospects of significant employment of non-inflationary nature in less developing countries. As a result, the exploitation of low-grade ores requires sophisticated techniques and is bound to involve movement of larger tonnages of material which generate large quantities of waste from sophisticated processing techniques. The impact of mining and mineral processing operations on the environment has thus grown to a great extent and will continue to have its aggregative impact as time goes on.

An attempt has been made in this paper to assess the ecological disturbances associated with the exploitation of low-grade ores in less developed countries and seek possible solutions. It may however be mentioned that it is rather difficult to differentiate between the effects as a result of high grade or low grade exploitation. The general impacts are same but vary in degree perhaps.

ECOLOGICAL DISTURBANCES DUE TO MINING

(a) Underground Mining

The removal of earth from the underground creates an obvious potential for ground movements and consequentional deformation of the surface.

This deformation of the surface is termed as 'subsidence'. Subsidence causes damage to surface structures, flora and fauna depending on the type and magnitude of ground movement. This in turn leads to ecological disturbances.

Control

To control the subsidence, special underground mining methods like harmonious mining should be adopted. The measures like special designs in the underground mining systems, adopting stowing of excavated area, rapid advancing of the working face and selecting a proper sequence of extraction, reduce the potential of surface subsidence to a great extent. However, where such practices are not possible, the surface is modified by bull-dozers and with filling of earth from neighbouring places to reduce the damage and make the surface even.

(b) Surface mining

Surface mining being the largest user of land usually creates greater problems than underground methods of mining. Undoubtedly, opencast mining causes significant degradation of the environment and to a large extent causes ecological disturbances as well. This is as a result of mining operations which entrail digging of the surface land. Many a times, jungles are cleared and agricultural fields excavated in the course of mining operations. Quite often such excavation leads to soil-erosion. The mine waste and overburden dumps are in many cases washed away into the nearby agricultural fields or water streams causing siltation problems. Some times, big landslides may be initiated by mining operations which may choke and temporarily or permanently change river courses. Such disturbances can be put together as land pollution which contains land degradation, landslides and soil erosion, etc. These can be considered as under.

Degradation of land

Usually surface mining methods produce large excavations especially in case of low-grade ore deposits, resulting in the spread of overburden and mineral wastes over large areas, making them unusable . The fine particles of the overburden and mineral wastes are carried away by the streams and wind and spread over agricultural land impairing its fertility. In some iron ore mines in Goa, which are low-grade iron ore mines, the overburden is dumped down the hillocks. These dumps are likely to cover paddy fields in due course of time making them infertile. In most of the clay mines of Kerala, the mineral wastes were dumped without proper planning damaging the agricultural land and causing deterioration of aesthetic values. Similarly, around big cities, the quarries of gravel and sand leave big scars which are ugly.

Landslides and erosion

Surface mining especially contour-mining, creates steep slopes along the highwalls, which may lead to rapid erosion and sometimes it may even cause land slides if not planned properly. In humid climates with steep slopes, a mass of water saturated soil overburden or weak clay or shale layers may slide down the slope in favour of earth floor and cause serious erosion. Mining creates a lot of landslide problems in hilly and other tectonically active regions where the ground is comparatively weak and unstable, thus disturbing the ecology of the region.

Control

1. A proper planning for land reclamation has to be done in advance of the extraction of mineral resources. As far as possible, the top soil should be preserved and should be laid over the abandoned area and reclaimed to make it useful for agricultural purposes.

2. Layout of dumps of waste rock have to be planned in such a way so as to minimise any environment.

3. Dumps of waste or barren rocks need to be utilized to such an extent that all useful components are derived to the maximum extent possible. Barren rock may be utilized as a cheap building material and as a filling material in underground mines.

4. Afforestation should be done in reclaimed abandoned mines and in barren areas. This may be done even in non-forest areas.

5. In the hilly terrains, especially in the Himalayas, a lesse should clear forest over a limited area only to start the mining operations. When completed, he may go to the next spot, clean the forest and start mining operation anew. While opening opening a new mine or extending an old mine, the area must be afforested.

II. Water Pollution

Mining operations cause water pollution through the discharge of acid mine water and by adding toxic substance like heavy metals, flourine and other radioactive elements. In deposits containing sulphide materials the water is transformed into acids and cause problems like corrosion of metal components, contamination of the water-table etc. This problem is especially serious while treating low-grade ores to extract the metal.

In dredging operations, the water pollution caused seems to be largely confined to silting. Generally these are associated with gold and china clay deposits. The effluents from the processing plants are generally added to the streams and rivers nearby which affect the people and the agriculture in the downstream side.

Control

1. Sealing of the source of mine water within the mines is an effective method.

2. Water re-use can greatly assist in prevention of water pollution.

3. Requirement and utility of mined land is an important pollution control technique which can bind pollution forming materials, reducing erosion and landslide.

4. Water treatment by various methods like, neutralisation, flocculation, precipitation, reduction, oxidation, biological processing etc. prevents pollution to a great extent.

III. Air Pollution

Exploitation of low-grade ores particularly cause air pollution by particulates and gases. Particulates result from the disintegration of solids and their suspension in the atmosphere causes pollution and ultimately leads to ecological disturbances.

The gases like carbon monoxide, sulphur dioxide, carbon dioxide produced in coal mining may cause health hazards to the workers. The refuse heaps and stockpiles in case of coal mines contaminate the air by emitting smoke and other gases due to spontaneous combustion.

Air is also polluted due to the dust produced during drilling, blasting, handling and crushing operations of large quantities of ore and waste, a common factor in the exploitation of low grade ores.

Control

1. During operations like drilling, blasting, handling and crushing, the process material should be sufficiently moist to suppress the dust and water is the usual medium. Installation of the spraying equipment at proper locations is necessary for effective suppression of dust.

2. Proper ventilation, increased recovery of coal and by employing drainage through the holes drilled, one can reduce the problem of fire hazard. Thus production of unwanted gases can be controlled.

IV. Noise Pollution

The unwanted sound is known as 'noise'. In mining it is produced by drilling, blasting, crushing and transportation. It causes a general feeling of annoyance, interference with unwanted sound and damage to auditory mechanism or reduction in efficiency of working performance.

Control

1. Noise control can be achieved by isolating the source, or hindering the transmission path or by improving the technology. However, a common precaution for a worker is to use ear plugs while on the job producing noise.

V. Ground Vibration

An undesirable result of explosive detonation is that the surface of the ground in the vicinity of the blast undergoes displacement. The vibrations produced due to blasting can cause damage to the surface structures, flora and fauna thus disturbing the ecology.

Air vibrations generated by blasting operations also causes damage to the structures, flora and fauna.

Control

The correct design of blasting rounds, avoiding the detonation of large unconfined changes and considering atmospheric conditions before blasting minimise the possibility of ground vibrations and air blast.

Some precautionary measures to be taken against ground vibrations due to blasting are :

1. Monitoring the vibrations from test-shots to determine site factors and the directional influence of local geology.

2. Inspection of buildings likely to be affected by blasting.

3. Careful supervision of drilling and blasting.

VI. Problems due to Mineral Processing

The life of a mineral processing plant has three distinct periods of environmental impact namely the construction period, the operating period and the period thereafter. There are three types of environmental problems which can arise due to mineral processing, especially in low-grade ore operations.

(a) Land disturbances

The effects of minerral processing, which play dominent role in extracting the metals from low-grade ore, operations resulting in land disturbances are:

- Aesthetic impact: due to mine excavations, spreading of overburden and mineral wastes, diversion of streams, deforestation, disappearance of wild and aquatic life, spoiling the scenic beauty of a place. Air and water pollution can also cause aesthetic nuisance by deflouring or turbunding streams by dust plumes etc.

- Degradation and decreasing the utility of the land, and

- Erosion of land by wind and water.

The dust from the tailings ponds creates major problems when it is blown by wind.

Control

- Stabilization techniques like vegetative, chemical and physical stabilization should be adopted.

(b) Air disturbances

Operations like crushing, grinding, screening, handling, dry concentration and dry tailing dumps create a lot of dust resulting in air pollution.

Control

- Water spraying systems for dust suppressions.

- Height of fall of the material should be reduced to prevent the formation of airborne dust.

- Velocity of airflow should be controlled.

(c) Water Regime

Mineral processing operations need large quantities of water. In contact with crushed ore, the water becomes highly contaminated by the unrecovered values of the ore and the waste which result in poor water quality, unsuitable for human consumption or even for industrial use. The polluted water in the tailings dam can also percolate through the ground, polluting aquifers.

Tailings

Generally in the processing of low-grade ores, a sizeable amount of tailings are produced. If they are not properly disposed, they may create a problem because the tailings may contain unwanted fine particles and even hazardous elements, etc.

Control

- Selection of proper site for tailings disposal, keeping in view the geology, use of land etc.

- Ensuring safe and stable foundations of tailing dams and preventing seepage of water.

VII. Ecological Impacts

Deforestation: In many cases the ore bodies of low-grade or high-grade are located in thickly forested areas. During the construction of infrastructure and pit site, the trees and vegetation are destroyed. In addition, the spreading of toxic overburden and dirt destroys the vegetation. Massive deforestation drastically changes the climatic conditions like rainfall, temperature and humidity.

The wild animals and birds which are living in the forest in a natural environment either die or desert the place due to mining operations. Sometimes the fish and other aquatic animals also die due to water pollution. Operations at Kudremukh Iron Ore India Ltd have affected the flora and fauna of the area severely.

It is reported that the different ecosystems and subsystems were significantly affected in Uttar Pradesh due to mining of magnesite, limestone, phosphorite and bauxite minerals at a number of places.

Control

- Planned reclamation should be carried out, and
- Afforestation should be done in reclaimed abandoned mines and in barren areas.

CONCLUSIONS

Ecological disturbances associated with the exploitation of low-grade ores have been examined. The changes in the environment due to mining affect the inhabitants directly and also indirectly through water, agriculture through biological byproducts and change of aesthetic values. The spectrum of ecological problems belong to not one discipline but is an interdisciplinary one connected with biology, geology, economics, engineering, sociology, etc.

In the light of the singular importance of the environment, the Govt of India and the Governments of less developed countries should evolve their respective national policies on environment. It is understood that the Govt of India is considering evolving a national policy on environment, the first step of which was the passing of "The Environmental (Protection) Act 1986" by the Parliament recently. Such measures are expected to protect man's environment from the ravages wrought by mineral exploitation, especially of low-grade ores.

MINERAL WASTES OF BIHAR - THE ENVIRONMENTAL IMPACTS

Akhileshwar Singh
Department of Forest and Environment (Bihar)
Patna, India.

ABSTRACT

Mineral processing wastes and manufacturing process wastes are hazardous products of mineral based activities and are threat to environment. The first recorded episode of hazardous mineral wastes release in India is probably that of Munger, in Bihar in 1968 when the river Ganga was set aflame. Bihar to-day produces mineral wastes faster than it can manage them. The State is presently producing thirty 'Major' and 'Minor' minerals, based on which there are hundreds of processing and manufacturing units all over the State. All these units are producing mineral wastes either in solid, liquid or in dust forms. The dust, effluent and leachates of mineral wastes are major pollutants, and the degradation of environment begins with generation and disposition of mineral wastes. Therefore, the desired pollution control measures need be treated as an integral component of mineral based industrial activities. The mineral wastes generated in the State of Bihar are identified, listed and analysed in the paper; and a view is advanced that most of the potential adverse effects could either be prevented or remedied, if mineral wastes are properly managed and reused.

INTRODUCTION

The Indian constitution recognises the need for environmental protection by incorporating it in its Directive Principles of State Policy. This aspect has been explicitly stated in Articles 51 A(g) and 48A. In addition, there are several Central and State enactments, of which the water (Prevention and Control of Pollution) Act, 1974, the Factories Act, 1948, the Mines and Minerals (Regulation and Development) Act, 1947, and Air (Prevention and Control of Pollution) Act, 1981 can be effectively used to control environmental pollution from mineral wastes. But there are several enforcement problems and inadequate appreciation of environmental problems. Consequently, adverse effects are glaring. Areas affected by generation of mineral wastes have a monochromatic appearance.

In Bihar, Dhanbad, Giridih, Hazaribagh, Palamau, part of S.P.(old) and Ranchi districts are depressingly black. Bokaro and Jamshedpur towns and the entire forest areas in the western part of Singhbhum are red. Wastes generated from limestone based industries have made the affected areas chalky white. There are ununiform layers of dust, on every available surface in the areas around stone crushers, numbering more than hundred in the s tate. All objects, including plants and animals in and around areas of intense mineral based activities, are curious looking.

In 1985, the Department of Science and Technology received a report from its working committee regarding pollution from mineral based industries. The report has identified the leachates and dust from mineral wastes as major sources of pollution. This is particularly insidious, since its role in polluting environment is enormous. In pursuance of the objective stated earlier and to counteract pollution of environment, a policy decision to introduce environmental aspects into mineral waste management need be taken explicitly and immediately.

MINERAL WASTES OF BIHAR

The State presently produces as many as thirty minerals and ores and imports about dozen such material for utilization in the State. These are listed in Tables 1 and 2.

TABLE 1

Minerals, Ores, and Rocks Produced in Bihar

Apatite	Copper ore.	Gravels and	Manganese
Asbestos	Dolomite.	Boulders.	Mica
Bauxite	Felspar	Garnet	Ochre
Bentonite	Fireclay	Graphite	Pyrite
Barite	Fluorspar	Hematite	Quartzite and
Ball clay	Graite	Kyanite	Silica sand
China clay	rocks.	Limestone	Salltpetre
Chromite		Magnesite	Steatite
		Magnetite	Slate
			Trap rock.

TABLE 2

Minerals, Ores and Other Raw Materials Imported into Bihar

Bauxite (high grade)	Gypsum
Barite (high grade)	Hematite (high grade)
Chinaclay (plastic)	Limestone (high grade)
Crude oil	Lead ore
Diaspore	Pyrophillite
Dolimite (high grade)	Rock Phosphate
Fluorspa, (high grade)	Silica sand
French chalk	Sulphur
	Sillimanite

The raw materials listed in Tables 1 and 2 are processed and utilised by about 250 small to large scale units, located in the State. Such units are clustered as well as unevenly located and can be grouped intomineral processing units and mineral-based manufacturing units.

The mineral processing units located in the State are -
Crushing and Grinding Units to pulverize ores, minerals and rocks at primary stage.
Plants to wash different type of ores, rocks and minerals,
Beneficiation Plants,
Blending Units,
Cutting and Polishing Units, and
Plants for treatment of mineral wastes like slag granulation unit, by-product recovery plants etc.

The number of such units, idle or working, are about 160, out which 80 are clustered around the Kiul-Shahaibganj-Pakur Loop Line, and between Kiul-Madhupur section of the Main Line of the Eastern Railway. Similar clusters can also be seen around Gaya and elsewhere in the State.

In Bihar, the Mineral-Based manufacturing units, both in the small and in the large sectors, are more than 90 as listed in Table 3.

TABLE-3

Existing Mineral-based Manufacturing Units of Bihar

Industry	No. of Units	Industry	No. of Units
Alloy	2	Lead metal	1
Alumina	1	Mica	4
Alumina ferric	2	Oil refinary	1
Asbestos Cement	1	Paint	1
Caustic Soda	1	Refractory	20
Ceramic	2	Steel igots	2
Coke	31	Sulphuric acid	2
Copper	1	Thermal power plant	8
Cement	8	Zinc Dust	1
Explosive	1		
Fertiliser	6		
Glass	3		

The minerals and ores of the State are, unfortunately, of marginal grade and significantly end up in waste dumps.

Mineral Wastes are generated during -

- the enrichment of ores, the evidence of which are available in the copper and iron ore belt of Singhbhum,
- the course of mining activities, the proof of which can be seen in South Bihar,
- the processing, manufacturing and extraction stages.

The mineral wastes of the last category is the subject matter of this paper and have been further classified into mineral processing wastes and manufacturing process wastes.

MINERAL PROCESSING WASTES

Disposal of the solid wastes obtained during mineral processing presents unique problems. It consists of crushing, grinding, simple washing and separation of valuless materials from desirable materials. The valuless materials vary from 10 per cent to 99.9 per cent of the R.O.M. This volume of wastes are being dumped, freely into the nearest valleys and where wastes are fine-grained, they form the source of dust. Wastes of all graphite processing plants of Palamau district are ultimately going into the North Koel river. Wastes from china clay washing plants of Singhbhum are accumulating into nearest agricultural land, valleys and ponds. The w ater of the South Karo and the Koina rivers are red. The same is the case with the Damodar valley. At Rajmahal, sandstone is washed to recover clean sand and wastes are being washed into the river Ganga.

The beds and banks of many streams, located near mineral processing plants, are coated with sticky brown and black precipitates. In the dry season, when discharge of most of the streams and the rivers becomes low, the river water becomes unusable.

All the crushing and grinding units of the State are freely polluting the environment. Breaking, crushing and grinding expands the volume of rock by 30 to 40 per cent and inthe process creates dust and noise. Heaps of mica wastes could be seen every where in the Bihar mica belt. Likewise, there are a dozen of coal washeries in the State, which altogether generate some 18 million tonnes of mineral wastes each year. All these mineral processing wastes are permanent source of pollution and contamination and these must be managed suitably to safeguard the environment.

MANUFACTURING PROCESS WASTES AND ASSOCIATED HAZARDS

The integrated steel plants, located in the State, presently produce 10.65 lakh tonnes mineral waste in the form of basic slag. Fly ash has wider impact on the environment. Presently, the thermal power plants, located in the State, together produce fly ash at the rate of 2400 tonnes per day. It has been estimated that by 1989-90, the production of fly ash will be one million tonnes per annum. Other Mineral based manufacturing units of the State, producing different kind of wastes, have been broadly grouped, analysed, and listed in Table 4.

TABLE-4

Manufacturing Process Wastes and Hazards

Mineral-based manufacturing units of Bihar	Manufacturing Process Waste's ingredients	Hazards
Oxidation and Reduction Plants Steel Plants, Copper and Lead smelters,	Carbon dioxide,Sulphur Sulphur dioxide,Basic Slag.Dust,Heat, Soot, Noise	Health hazard. Damage to vegetation. Water contamination. Atmospheric Pollution
Thermal Power Plants and coke ovens	Sulphur, Titanium, Sulphur dioxide,Carbon dioxide,Fumes,Heat Dust,Nitrogen oxide, Fly Ash,Volatiles.	Soil acidity. Air Pollution, Water contamination. Damage to plant.Health Hazard. Smog.
Cement Plants	Dust,Sulphur dioxide, Dust Cake.	Abrasion, Water contamination,Damage to plant and soil fertility. Air pollution. Health Hazard.
Asbestos Cement Unit	Fibre dust.	Asbestosis.
Alumina Plant	Red mud (Contain iron, vanadium and titanium)	Toxic effluents.
Caustic Soda Plant	Chlorine,Mercury, Hydrochloric acid.	Water contamination.

Glass Plant	Fluorine	Fluorosis
Fertiliser Plants including Sulphuric Acid Plants	Sulphur dioxide, Fluorine, Amonia, Arsenic, Chromate.	Allergies. Toxic effects. Problems with tooth. Acid rain. Water contamination. Air pollution. Bronchial Problems. Damage to kidney.
Oil refinary	Mutagenic and Carcinogenic chemicals	Water contamination. Serious health hazard.

SUGGESTIONS TO ALLEVIATE THE ENVIRONMENTAL IMPACTS OF MINERAL WASTES

The mineral-based activities have very important role in economic and industrial development. In the financial year 1985-86, the State collected Rs.3600 millions as royalty and cess on minerals. It will be trite to emphasise the pervasiveness of mineral-based activities. Since the generation of mineral wastes cannot be stopped the environment can be only protected by controlling generation of mineral wastes and by suitable management. Mineral wastes consist of many valuable products. They can be recycled and put to use to extract the valuables. Hence, impact of mineral wastes on environment can be sufficiently reduced. More studies are needed in this regard. In this context, the suggestions and the possible uses of mineral wastes have been listed in the Table 5.

TABLE-5

Measures to Reduce Environmental Impacts From Mineral Wastes

Mineral Wastes of Bihar	Possible uses and suggestions
Red mud of Alumina Plant	Processed liquor contains both vanadium pentoxide and titanium oxide. The titanium oxide percentage goes upto 25% and this should be recovered.
Copper furnace slag, leachates of beneficiation plant and by products of copper smelter.	Gold, silver, selenium, nickel sulphate, and tellurium are being recovered as by-products. There is need to recover cobalt, molybdenum and magnetite. Copper furnace slag can be used as cheap additive for consolidating stowing materials. Also recovery of more copper by acid and bacterial leaching should be experimented. R.R.L, Bhubaneshwar has developed a improved process for extraction of metal.
Basic slag of integrated steel plants.	Some by-products like coke oven gas, tar, etc. are being recycled. In the State, the capacity of slag grinding units should be increased and use of slag as a soil amendment

	and phosphatic fertiliser, should be promoted. Slag produced in the State contains 43 to 52% Cao. Therefore mixing higher proportion of slag in cement klinker should be examined Use of slag as construction material should be promoted.
Fly Ash of thermal power plants	Use as pozzolana and stowing materials should be promoted. Technology is available to manufacture bricks from fly ash which is cheaper than conventional bricks.
Wastes from coal washery including coal ash, middlings, washery sinks etc.	Use in power plants by "Turn Key Technology import" be promoted. The by-product plant study like recovery of titanium, rare earths, manufacture of hollow bricks, and use as paving material should be promoted. Use as blend in phosphatic fertiliser, and in the construction of dykes may be examined. Rejects and middlings industries including fuel briqutte plants should be promoted.
Wastes of beehive coke plants	Recyling of gas for heat generation and recovery of volatiles including tar product, benezole etc. are possible.
Fine conoentrates generation as a result of beneficiation of Kyanite	In high alumina refractory, kyanite used is in the form of finer fraction. This is being preferred in developed countries.
Wastes generated by existing graphite beneficiation plants.	A cheaper and better beneficiation method should be evolved. Presently, the percentage recovery is only 15% as against 90.8% reported in N.M.L. pilot plant testing.
Wastes generated by china clay washeries.	Only 15 to 30% of the crude washed clay is obtained as refined clay. Washing procedure should be improved to reduce generation of rejects.
Wastes of beneficiation plant located at Rajmahal, S.P.	Recovery and washing should be improved to reduce generation of wastes.
Wastes of limestone beneficiation plant.	Considering the lower fuel consumption and the lower requirement of water and better recovery, the dry process should be preferred.
Mica scrap and wastes (by faulty processing).	Improvement in processing techniques, export and use in mica paper, mica bricks, and mica powder industries should be promoted.
Ferrous manganese slag.	Preparation of manganous sulphate monohydrate is now possible.

More studies are needed on the subject of utilisation of mineral wastes. Technological innovations in existing mineral processing techniques and in mineral-based manufacturing process are necessary to regulate and curtail generation of mineral wastes. In fact, this is a necessity if we have to conserve mineral resources and to protect the environment.

MANAGEMENT OF MINERAL WASTES - AN IMPERATIVE

We have seen that the most serious environmental problems of the mineral resources industries relate to mineral wastes. Furthermore, they represent wasted resources. If these could be recycled or valuables are recovered, the demand on the minerals resources industries would be lessened.

Mineral wastes are also potent causes of solid pollution, liquid pollution, gaseous pollution, biological pollution and visual pollution. We must readjust our thinking about mineral wastes which should be viewed as renewable resources, returnable to the economic cycle. To accomplish these objectives, integrated recovery systems should be developed, incorporating the best features of many processes. Besides being a mineral conservation measure, the development and use of a comprehensive ore processing system would have significant environmental benefits.

ADMINISTRATIVE AND FISCAL MEASURES

Conservation of mineral resources and the preservation of the environmental quality are important objectives. However, the situation is not likely to improve unless 'benefit-cost' criterion is reviewed and the environmental quality incorporated into it to judge the real worth of mineral based projects. There is urgent need for levying an 'Environmental Development CessS' on the basis of the quantum of mineral wastes generated. The State Department of Mines and Geology should recycle 10% of the amount collected as royalty and cess on minerals, for development of environmental quality. Tax incentives or subsidies might be granted to encourage the development of better recovery systems and the best features of many processes. In order to effectively manage the mineral wastes generated in the State, the State Government may set-up a "Mineral Wastes Management Board" under its Department of Forest and Environment.

REFERENCES

Centre for Science and Environment. Second Citizen's Report on the State of India's environment (1985)

Howard,A., Geology in Environmental Planning, McGraw Hill, New York, Hill, U.S.A. (1978)

Kalra,G.D., Mineral Development in Bihar, published by NCAER (1983)

Khoshoo,T.N., 27th Holland Memorial Lecture on Land Reclamation in Opencast Mines (1984)

Sinha, Basawan, Assessment and valuation of Natural Systems and Environmental consequences of Water Ressources Development (1985)

* Singh, Akhileshwar, Need for utilization of leaner and unmarketable ores in Bihar. Seminar on Mineral Processing, N.M.L.,Jamshedpur (1985)

* Necessity of particle size studies of Bihar's ore by the Mineral Industry. Seminar or Mineral Processing, N.M.L. Jamshedpur (1985)
* Search for Technology to maximise pyrite utility in Bihar. Seminar on Mineral Processing, N.M.L.,Jamshedpur.(1985)

* Bihar's mineral in Small Deposits - Its present and future. Regional (Eastern Zone) Seminar on Small Mineral Deposits, Ranchi.(1985).

* Role of Private and Public Sector in Development of Bihar's Mineral Resources. Workshop on Prospects of Industrialisation in Bihar. Ranchi (1985).

* Technological innovation for utilisation of Marginal Grade Limestone International Seminar on Cement and Allied Building Materials, Delhi (1987). To be published.

SOME PROBLEM DIMENSIONS

PROSPECTS OF MINING SMALL SCALE BASE METAL DEPOSITS IN INDIA

A. N. Bose
Indian Bureau of Mines, Hyderabad, India.

ABSTRACT

Base metals play a vital role in the development of a nation and India is deficient in these metals. There is a wide gap between the growing demand and indigenous production which is being met by import. Even though a few large base metal deposits are currently under active exploitation/development, an increasing need has beenf elt for mining small and medium scale deposits with reserves varying from a couple of million tonnes or less to about 5 million tonnes, capable of yielding a production of say upto 500 tpd. While few such deposits are being mined or have been mined in the recent past, scope exists for developing at least a few more promising prospects into producing mines. Working of these deposits will also generate employment and help in the balanced growth of the country. Problems of the existing small base metal mines in the country, the quantum of exploration/exploratory mining carried out in these promising prospects, ore reserves and feasibility of developing them for commercial exploitation have been discussed and certain remedial measures suggested. The role of Indian Bureau of Mines indeveloping these small scale deposits has also been outlined.

INTRODUCTION

Base metals play a vital role in the service of mankind. These metals find extensive use in almost every sector of the economy such as energy, transport, communication etc. The likely demand and anticipated production for base metals in the country have been furnished in Table - 1 below:

Table - I

Likely Demand and Anticipated Production of Base Metals in India

Tonnes

Item	Approximate demand by the end of (1984-85)	Production achieved by 1984-85	Demand projections for the end of VII plan (1989-90)
Lead(primary)	56,000	24,600	69,900
Zinc	1,44,600	68,500	1,80,200
Copper (Refined)	1,17,500	50,000 (approx)	1,57,300

It is thus clear that there is a wide gap between demand and indigenous production warranting exploitation of small or medium scale base metal deposits even though they, by themselves, cannot sustain economic metal production facilities. In fact, exploration work for base metals by the Geological Survey of India (GSI) and other agencies have revealed the presence of a large number of small and medium scale deposits all over the country in addition to a small number of large deposits, almost all of which are currently under exploitation. The latest find is at Rampura Agucha in Rajasthan State where substantial additional reserves of high grade zinc-lead ore have been proved and the deposit is being developed for regular production. In spite of this, it may not be possible to attain or approach self-sufficiency in the production of lead and zinc and much less so in respect of copper. This gap can, somewhat, be narrowed by mining smaller deposits thus utilizing the existing reserves to the full extent, even though many of them have been hitherto neglected because of economic reasons.

In the Indian context, a small base metal deposit has been defined as one which can sustain an annual production upto 25000 tonnes with 66% recovery over a life of 20 years and the figure is doubled to arrive at the mineable reserves (i.e. 1.5 million tonnes) which should be available in the deposit to take care of its likely expansion to a somewhat bigger mine. Total geological reserve of possible category is again double of this figure i.e. 3 million tonnes in small deposits. In India, some of the small scale deposits have been developed and worked by the public sector undertakings specially formed for the purpose by the respective States under the Indian Union. Occasionally, there has been participation/financial assistance of the Central Government also in such ventures.

PRESENT SCENARIO

General

Even though a very large number of small lead, zinc, copper deposits have been located in various parts of the country, only a few have so far been taken up for active exploitation and still fewer could reach their designed capacities. Mining operations in a couple of them had to be discontinued as they were not found to be economically viable.

Small Scale deposits under exploitation

Six such deposits are currently under active exploitation in the country. Mining operations at Bhotang mines in the border State of Sikkim date back to 1890. The salient details in respect of these deposits, present mine capacities, targetted/actual annual production of concentrates or metals, etc. are furnished in the following table:

Table - 2

Salient details of small case base metal deposits currently under exploitation in India

Deposit/ mine	Metal reser- ves in lakh tonnes. (Av.g (Av.grade in paren- thesis)	Mineable daily ore produc- tion (tpd)	Av. grade of ore produ- ction	Av. grade of mined	Produc ction of conc. (with Gr) (tpy)	Remarks
Agnigundala (Bandalamottu)	Lead	20.00 (5.07% pb)	240	4.26% pb	4538 (66.4% pb)	-
Ingaldahl	Copper	14.00 (1.34% Cu)	135	1.12% Cu	1233 (N.A)	Figures upto Nov.'85
Kalyadi	Copper	30.40 (0.88% Cu)	60	1.02% Cu	370 (22.5% Cu)	-
Mailaram	Copper	4.86 (1.45% Cu)	110	1.2 to 1.5% Cu	1800 (22 to 25% Cu)	Figures suggested by IBM TC Cell; Concentrator still under trial runs.
Dariba	Copper	4.00	100	1.45% Cu	2000 (22% Cu)	Recent figures not available.
Bhotang	Copper, Lead, Zinc	1.2 (1.13% Cu) 1.60% Pb, &3.21% Zn)	60	0.79%Cu 1.40%Pb 3.31%Zn	470 (19.5% Cu) 101 (40.57%Pb) 528 (33.18%Zn)	There is no definite cut off.

* 1 lakh = 0.1 million

It may thus be observed from the foregoing table that proved/demonstrated reserves in all these deposits are rather low and so is the grade, for which development of these mines with larger production capacities is not feasible, both technically and otherwise. The operations tend to be uneconomic because of these constraints. However, the plicy envisaged for utilisation of these deposits is one of developing them in such a manner that they can serve as satellites or feeders for centrally located smelters. These deposits, in fact, have been opened up by development of mines and setting up of concentrators and the concentrates produced are being transported to smelters owned by the large public undertakings under the Central Government for metal production.

Nature of mining operations in brief.

Since mining has been mostly carried out by underground methods, shallow vertical shafts have been sunk as modes of entry. In a couple of cases, however, adits/inclines have formed as the modes of entry to suit the nature of occurrence in particular cases. They are usually small diameter shafts (average 3 m) which are used for ore hoisting as well as for transport of men and materials.

The method adopted for the development of the deposits is through cross-cuts and by level drivage usually 2.5 m x 2.5 m in cross section. They are then connected by raises and winzes with similar cross sections to block out the ore. Compressed air drilling with jack hammers and air legs is the common practice using either "burn cut" or "wedge cut" pattern of drilling in drives and cross cuts. The number of holes per round varies from 24 to 30 and the advance per round is about 1.2 m. Special gelatine (60 to 80% strength is used for blasting using half-second or millisecond delay detonators.

The methods of stoping usually adopted in these mines are either "shrinkage" or "sub-level" method without filling. Production capacity of a stope is usually around 120 tonnes per day. Apart from leaving a crown pillar of about 3 m. and a sill pillar (5m) at the bottom of the stope, protective pillars are left on both sides of the service raises kfor stability of acess ways during the life of a stope. While the mining losses are of the order of 20-25%, dilution from wall rocks and intermediate partings is of the order of 10% or slightly more. The mine management are required to obtain prior permission of the Indian Bureau of Mines for commencement of stoping under Mineral Conservation and Development Rules, 1958. The stoping proposals are carefully examined from the angle of conservation of minerals, adequacy of exploration, etc. before kpermission is granted.

Loading of tipping tubs of 1 to 1.25 tonnes capacity underground is usually done manually and occasionally by rocker shovels on rails. The loaded tubs are hand-trammed and the skips are usually loaded through manually operated chutes. In respect of the mines served by adits, the ore/waste is hauled up to the adit level and then hauled out of the adits by locomotives. The ore is then transported to concentrators.

The requirement of mining personnel for a copper mining project was worked out as 286 for 250 tpd capacity and 506 for 500 tpd capacity. While the number of skilled workers will be 32 in the former case and 58 in the latter, that of semi-skilled 92 and 132 respectively and the remaining are unskilled workers.

Problems of base metal mining in India.

Mining of small scale base metal deposits in the country which has contributed its share to bridge the gap between the increasing demand and domestic production has, however, suffered from a number of limitations and problems which are discussed below:-

i) Although these mines have been developed and operated by public sector undertakings or their subsidiaries owned by different States or the Indian Union, shortage of funds have often been experienced.

ii) Being small units, there has always been a lack of proper infrastructural and other supporting facilities.

iii) Lack of proper mine evaluation and planning. Exploration efforts to locate new deposits inthe neighbourhood have also been limited.

iv) By and large, the grade of the ore available for mining has been rather low.

v) Lack of experienced and properly trained personnel at various levels for operating these mines at optimum level of efficiency.

vi) High transport cost of concentrates to the smelters of large producers.

vii) The price of concentrates is dictated by the large producers with little or no consideration for the cost of production in individual units from where the concentrates are purchased for smelting. This is a key factor which determines the profitability of a small base metal project in India.

viii) Concentrators have also not functioned at the desired level of efficiency.

ix) The quality of concentrates has not been satisfactory owing to contamination by other metals and there has been considerable loss of metals in the tailings in some cases.

The last four factors are significant so far as the small scale base metal mining operations are concerned. Barring exceptions, concentrates produced at a small base metal mine necessarily have to be transported at present over long distances to the smelters. Thus, the price realisationat the mill head per tonne of concentrate (the concentrator is usually located at the mines) is diminished by the figure for freight and handling charges. To cite an instance, for copper concentrates which were bought inther ecent past by the Hindustan Copper Ltd., the sole copper producer in the country, from a small copper project in South India the price presently being offered per tonne of concentrate (25% Cu. content) is around Rs.9,000/- (about US $720) at Ghatsila in Eastern India. The corresponding railwayfreight alone per tonne of concentrate is Rs.312/- (approx.US $ 25). It is true that over the years, the prices being offered by such large producers have also increased progressively with cost escalation. Yet, it has failed to make exploitation of many of these small deposits viable and most of them have incurred recurring losses resulting in mining operations on a couple of them being discontinued as mentioned earlier. As regards the problems mentioned at (ii) and (iii), in the preceding paragraphs, the Indian Bureau of Mines have provided necessary guidance and assistance to small base metal producers in various ways. In the case of a State Mining Undertaking, the Technical Consultancy Service of the department was availed by public financial institutions for a financial-cum-technical appraisal of a small copper mining project for financial assistance which had considerable impact on the execution of the project. In the case of yet another multi-metal project, a similar exercise was done by the IBM for processing a financial assistance. Mining Engineers, geologists and other specialists of the IBM have carried out geological appraisal/detailed exploration of some of these deposits. Preparation of detailed mining schemes, equipment selection, etc. have also been carried out by the IBM mining engineers. These professionals have also rendered valuable advice from time to time for systematic development of the mines consequent to their inspection of these mines and for optimising their utilisation and have, in fact, compensated to some extent the lack of properly trained technical personnel at these mines.

Leaching has found wide application in the recovery of copper using weak acid solutions to dissolve copper in oxidised ores. It is relativelycheap to introduce since it requires little in the way of sophisticated equipment. The low capitalcost of the process coupled with a short construction period could act as an incentive for development of small mines. Some work on bacterial leaching has also been done in India and this technique could be commercialised for recovery of copper from low grade material. This could also contribute significantly to the success of new small copper mining ventures.

PROSPECTIVE SMALL SCALE DEPOSITS FOR FUTURE EXPLORATION

Deposits exploited earlier where mining operations have been suspended.

Two such deposits fall in this category - one for zinc & copper and the other for copper. The Deri mine in Rajasthan for zinc & copper was worked at first by the State Department of Mines & Geology and later by the State owned Rajasthan Mineral Development Corporation (RSMDC) on a very small scale. The reserves were estimated to be of the order of 1.0 million tonnes with 5.87% Zn; 4.4% Pb and 1.98% Cu. The mine has been developed through two vertical shafts (2.5m x 4m) to depths of 29 and 33 metres respectively. A single level has only been developed approximately at a depth of 30 metres below surface. Three out of the four mineralised zones (revealed by surface exploration) have been partly exposed. A techno-economic feasibility report for the mine has been prepared by the Hindustan Zinc Ltd (HZL) at the instance of RSMDC which is reportedly under examination.

Similarly, a small copper mine at Thinthini in the Gulbarga district of Karnataka State had to be closed down since early 1979 as the project was not considered viable. This mine was developed through three shafts and a system of cross-cuts, drives, raises and winzes. About 36,000 tonnes of ore were treated and about 500 tonnes of copper concentrates analysing on an average about 24.4% Cu. were produced from this mine prior to discontinuance of mining operations.

Virgin small scale deposits explored in fair detail but where regular mining operations/concentrates production have not yet commenced.

Ambamata multi-metal deposits in Gujarat State.

A number of agencies carried out exploration of the deposit during different periods commencing from 1966. Limited underground development has also been done. The deposit could thus be considered as reasonably well explored. Close spaced drilling on 30 m grids had been done in the northern ore body particularly for the area earmarked for open pit mining (both open pit and underground methods are contemplated). A "pilot" open pit has already been opened up in the demarcated area. 10 m high benches in the overburden and 5 m high in ore are proposed in open pit operations to prevent ore dilution in the latter case with a stripping ratio of 8.4 : 1 by weight. The mining reserves have been updated by a renowned group of consultants as follows:

a) Open pit - 2.1 million tonnes at 3.00% Pb, 5.33% Zn and 1.2% Cu.

b) Underground - 5.3 million tonnes at 2.9% Pb,4.79% Zn, and 1.5% Cu.

While mining of ore (targetted throughput : 3.5 lakh tonnes per annum) is not likely to pose any problems, the real difficulty lies in the production of sufficiently clean concentrates of each of these three metals viz. lead, zinc, and copper. Total metal recoveries have also been found to be low. On the other hand, production of bulk sulphide concentrates resulted in a much higher total metal recovery. Recovery of individual metals from the bulk concentrates is an established process and it is felt that this could be done by toll smelting abroad estimated to cost $ 90-100 per tonne. A firm decision in this regard is yet to be taken. The availability of adequate finance for pushing the project ahead is also a severe constraint in this particular case. The project has remained practically dormant for the past 2 or 3 years. Once these problems are solved and decisions taken, the deposit will contribute significantly to the base metal production in the country as the targetted production of bulk sulphide concentrates from this deposit will be around 65,000 tonnes at full capacity.

A 50 tpd pilot plant was commissioned in the project site in February, 1980 costing about Rs.40 lakhs (US $ 3,20,000), 50 percent of which was financed by the Ministry of Steel & Mines of the Government of India as a R&D project. But due to acute shortage of water, it has been possible to run the plant only intermittently. Some 80 tonnes of bulk concentrates were produced from this pilot plant till January, 1981, by treating the ore obtained from the pilot open pit and from an incline driven from this pit. The concentrates produced analysed (with about 25% recovery) as follows:

Pb	:	12 to 15%
Zn	:	20 to 25%
Cu	:	4 to 6%

Dikchu Copper Zinc deposit in Sikkim

The massive Dikchu ore body occurs as a single sheet of mineralised unit sandwiched in mica schists. The total strike length is about 670 m. The sulphide ore body is practically copper-zinc mineralisation. The exploration wing of the Indian Bureau of Mines carried out detailed exploration by exploratory drilling and a level development of the Dikchu lode way back during 1959-63. Apart from detailed geological mapping, exploratory core drilling in 17 bore holes totalled over 4600 m and exploratory mine development in 4 adits added upto a little over 1500 m. This work was followed by further investigation by Mineral Exploration Corporation Ltd. 1950m of underground development was planned, about 95% of which was completed by March, 1982. The following tentative reserves were estimated bdy the MECL:

i) Mine developed reserves - 2,41,260 t.
ii) Proved reserves - 97,020 t

3,38,280 t

As regards the quality of the Dikchu ore, three representative bulk samples weighing about 500 kgs. were collected and one of them intended as average grade sample yielded very high Cu. and Zn values. The expected values are around 3% Cu. and 1.5% Zn. Even though mining conditions in the Himalayan region are quite difficult, this deposit holds promise for future commercial exploitation particularly in view of the nearby Bhotang mines which is a small producing mine of the Sikkim Mining Corporation, a joint venture of the Governments of India and Sikkim.

Mamandur Multi-metal Deposit in Tamilnadu

Base metal mineralisation at Mamandur is localised along a shear zone at the contact of garnet sillimanite gneis. Intensive geological, geophysical and geochemical exploration followed by exploratory mining (3 levels) revealed that it is basically a zinc prospect with lead and small amounts of copper with an aggregate metal content of 5% with silver (20-260 gms. per tonne) in galena and cadmium ranging upto 0.25% in sphalerite. Persistence of the mineralisation has been traced to below the 200 m. level from the surface. Reserves have been estimated by the UNDP in the range of 1 to 1.5 milliontonnes. Silver and a cadmium values form extra incentives for de veloping this small scale prospect into a producing mine. The State owned Bharat Gold Mines Ltd. (BGML) had examined the techno-economic feasibility (for which too the services of the IBM were retained) of its commercial exploitation. It appears that the deposit is not workable under the present economic conditions and work in this deposit was abandoned way back in January,'79.

Promising small scale prospects investigated or under investigation which could be converted into producing mines.

A few small base metal prospects are currently under investigation by the Hindustan Zinc Ltd., the main producer of zinc and lead in the country. They include Gorubathan, Devpura, Dewdas, Sindeswarkalam West & East etc. The demonstrate d reserves (ie. proved and probable reserves taken together) in the first four prospects named above add upto about 8 million tonnes of which Dewadas appear to be quite promising with a demonstrated reserves of 3.65 million tonnes and a possible reserve of further 2.16 million tonnes, average values being 2.61% Zn and 1.50% Pb. Quality-wise, however, the Gorubathan deposit (demonstrated reserves 1.61 million tonnes) in West Bewngal appears to be the best (4.03% Zn; 3.95% Pb). Salient details in respect of certain promising base metal prospects in the country and the feasibility of developing them into small producing mines have been furnished case-wise in the following paragraphs:-

Zangamarajupalle Lead Zinc project

Considerable exploration has been carried out at Zangamarajupalle Lead Zinc deposit in the Cuddapah district in the State of Andhra Pradesh in South India. A total of 58 boreholes aggregating about 7,700 m. have been drilled which brought to light this small scale base metal deposit. The demonstrated reserves are about 2.19 million tonnes with an average grade of 1.12% Zn and 3.02% Pb. A usual cut off of 3 percent Pb + Zn is likely to yield a possible commercial average grade of about 5 to 5.5 percent Pb + Zn. Subject to exploration mining, it is reasonable to expect a modest production of 200 to 250 tonnes per day with around 5 to 5.5 percent total metal content from this deposit, over a period of about 20-25 years. About 4 to 5 percent total metal content could be considered as economic grade at present.

Nallakonda copper prospect

Nallakonda Copper prospect (Agnigundala belt) in the State of Andhra Pradesh was first explored by the Geological Survey of India (about 15,000 metres of drilling in 72 boreholes) followed by the Hindustan Copper Ltd. (about 850 metres of exploratory mine development in 4 levels and about 490 m of underground drilling). A mine with a small production of 100 to 150 tpd can be built at this deposit to last for 10 years provided the entire mineralised zone is taken up for development.

Basantgarh Copper Zinc prospect in Rajasthan State

Promising Copper Zinc mineralisation has been encountered at Basantgarh in Sirohi district of Rajasthan. "Drill indicated" ore reserves are of the order of 3.5 million tonnes, 1.74% Cu. and 1.27% Zn. At the instance of Hindustan Copper Ltd., infill drilling was carried out by the State Directorate of Mines & Geology so that exploratory mining could be taken up and bulk samples collected for beneficiation tests. Pilot plant trials on mineralised cores carried out in the IBM laboratory revealed the feasibility of obtaining a concentrate analysing 18.8% Cu and 1.48% Zn with about 91% copper recovery.

Akwali Copper prospect in Rajasthan

Detailed exploration by drilling (9200 m) and 2090 m of exploratory mining (one level development) has revealed the existence of a lode about 700 m in length with anaverage width of 3m. Estimated ore reserves (indicated & inferred taken together) between 620 - 250 MRL with 0.4% Cu. as cut off totalled 1.65 milliontonnes while that at 0.8% cut off between the same levels totalled 0.98 million tonnes. There are fair chances of developing this small scale deposit into a small mine to sustain a daily production of about 200 to 250 tpd which could also augment the supply of ore to the Khetri Smelter of the HCL. Besides copper, gold (0.5 to 0.6 gm/tonne), silver (6.0 to 8.0 gm/tonne), Zn (small zones assaying between 1.15% to 5.70%) and cobalt may be recoverable as by-products.

Bhagoni Copper prospect in Rajasthan

It is yet another promising copper prospect situated in the Alwar district of Rajasthan. The ore zone is about 7 m. wide over a strike length of 1200 m. Both surface and underground drilling totalling over 16,000 m. in 69 boreholes have been done followed by a total 1837 m. of exploratory mining (one level underground development). At 0.5% cut off, reserves (indicated + inferred) + totalled 5.2 million tonnes averaging 1.07% Cu. This includes 0.35 million tonnes of 1.22% Cu. from exploratory mine development.

Subject to further work, feasibility of developing this deposit into a small mine upto 500 tpd capacity could be examined in detail.

CONCLUSIONS

While half a dozen small scale base metal deposits are being mined in India, there are a few promising prospects in North-East, West and South India which could be taken up for exploitation on a small scale to augment base metal production in which the country is deficient.

Instead of operating the small base metal deposits by a number of State Public Sector undertakings individually with limited resources, it may be advisable to entrust the work of rapid development and exploitation of these small scale deposits in an integrated manner to a separate public sector agency for optimising development of mines, deployment of personnel, production and pricing of concentrates related to the cost of production, etc. This would be an effective way of promoting exploitation of small scale deposits.

Exploration efforts in the neighbourhood may have to be intensified to locate any new deposits which, by itself, may not be amenable for commercial exploitation. One such deposit (2.2 lakh tonnes with 1.46% Cu) has been located close to a small working mine and a scheme for its exploitation has been drawn up.

Feasibility of reopening small closed mines may have to be examined in depth as some development has already been carried out and infrastructural facilities are available to some extent. Judicious grouping of small deposits for their collective exploitation also deserves serious consideration.

Even though some of the R&D programmes for hydrometallurgical recovery have received active encouragement from the Central and State Governments, these efforts need stepping up both in respect of metallurgical aspects as well as for development of improved mining techniques.

R&D efforts need be intensified in the field of bacterial leaching which can be conveniently used for concentration of low grade copper ores usually found in most of the small sized copper deposits in the country.

ACKNOWLEDGEMENT

The author is grateful to Shri D.N.Bhargava, Controller General, Indian Bureau of Mines, for the encouragement received from him and for his kind permission to present this paper. The author wishes to thank his colleagues in the department, and the officers of the sister departments/undertakings like the GSI, MECL, HZL, etc. whose reports were freely consulted and for their valuable suggestions during the preparation of this paper. However, the views expressed inthe paper are those of the author and not necessarily of the department.

REFERENCES

1. Bhargava, D.N. : Report on International Conference on small scale Mining of the World, Mexico from 27 Nov. to 5 Dec.,1978.

2. Bhatnagar, S.N. & Gandhi,S.M. : "Status of Lead, Zince Resources & Future Perspective"- International Seminar: Lead, Zince and Cadmium, New Delhi, Jan, 1981.

3. Controller General, : Proc. of the workshop on Mineral Policy for Small Scale Mining, July, 1985.

4. Ganapati, R. : 'Non ferrous metals in Indian - perspective of growth' - J. of Mines, Met. & Fuels, Jan.,1983.

5. Government of India Ministry of Steel & Mines Department of Mines. : Report of the Working Group on Non Ferrous Metals (Lead and Zinc) July, 1984.

6. Mining Journal Ltd. : Mining Journal Vol.306, No.7867 p.385.

7. Raghu Nandan,K.R. Dhruwa Rao,B.K. & Singhal,M.L. : Exploration for copper, Lead & Zinc ores in India - Bul.of GSI- Series A -Eco.Geology No.47 (1981).

LIFE INDEX OF MINERAL RESOURCE - CASE FOR COPPER ORE IN INDIA

S C Ray
Department of Mining Engineering
Indian Institute of Technology
Kharagpur, India

Balbir Singh
Hindustan Copper Ltd
Calcutta, India

S K Mukhopadhyay
Department of Mining Engineering
Indian Institute of Technology
Kharagpur, India

ABSTRACT

In recent years the adequacy of some mineral resources to meet the demands anticipated by future growth has become a problem of serious concern. A few of the important factors responsible for the uncertainities associated with long term supply demand balance are that, resources already known to exist will be supplemented by new additions, technological change will affect extraction cost and substitution among resources will result from changes in relative prices.

Life index concept arising out of production capacity is not meaningful because, for the short term, shortages are generally due not so much to low reserves as to inflexibility of production capacity in responding to rapid swings in demand and for the long term, due to the essential question of future supply should be addressed not to currently proved reserves but to our ability to develop additional reserves for desired production levels.

Likely pattern in which the current reserves will be mined out can be projected but all the controlling factors like size of reserve, generation of new reserves, production capacity cannot be varied as well as they are restricted by geological and economic factors. With the data of growth rate and resource use, the time of exhaustion of a resource can be calculated mathematically. Discussing the general background of life index and resource availability the position of copper ore in India and its ability to sustain demand is discussed. The influence of secondary supply is also incorporated.

INTRODUCTION

For the present day industrial siciety a regular supply of mineral resources is essential. In the past, countries did not concern much with long term availability of resources primarily because the amount of mineral resources in the earth appeared to be nearly infinite as well as the annual usage rate were not much.

In the last decade and half, two trends have changed this picture. The usage rate of mineral resources has been growing at an exponential rate. Also countries are beginning to expand their planning horizons. These two trends lead to the realization that the mineral resource supply will be critical in future and hence the concept of life index of mineral resource has gained in importance.

For the purpose of regional distribution of production and world's mineral resources, world was divided into industrialized nations, developing countries and nations with state economics. It has lately been found that the availability of mineral resources is more critical when one considers the resources within one national boundary. This problem is not so critical with developing and less developed countrieds. Nevertheless, they are also concerned as the industrial growth is increasing leading to increased demand for mineral supply.

LIFE INDEX CONCEPT

The need for a measure of resource availability has led to the use of some measure, namely the life index, which indicates the number of years current known reserves could support consumption at its present level. It is a misleading statistic which is used as a measure of adequacy of a mineral commodity on a national scale. This concept may be taken for a single mine but the extension to several dissimilar mines is not justified. In Fig. 1(a) the life index of a deposit is shown whereas the life index of the same mineral will be much less if we take into account the growth rate in production, Fig. 1(b). However, it is true that total

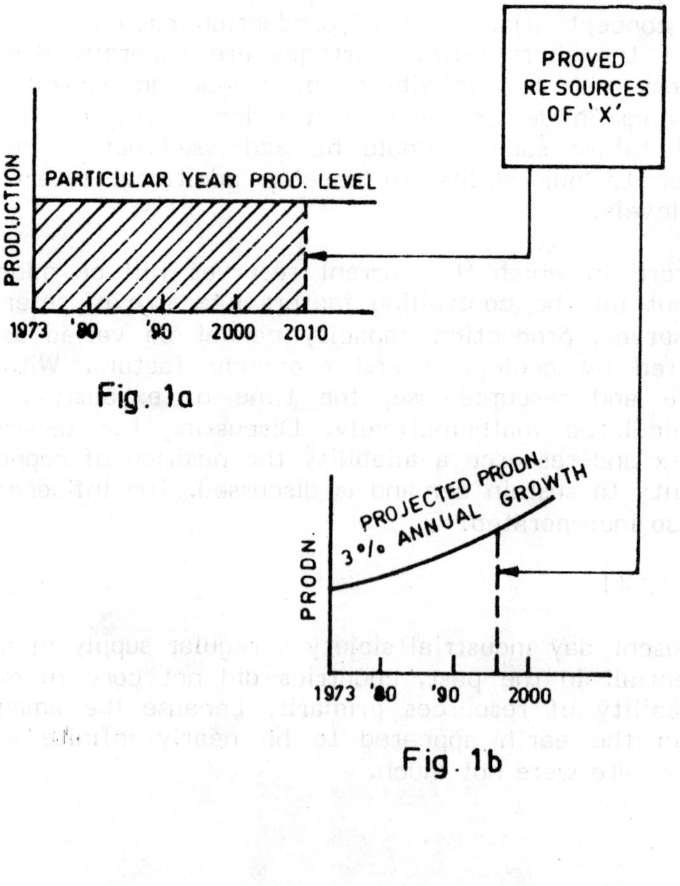

Fig. 1a

Fig. 1b

production in future will not come from the currently known deposits alone as taken in the figure. Some mines will be exhausted whereas new mines will come up that are not yet thought of. Thus the measure of life index, while useful as a very rough approximation to resource availability, does not reflect the complex interactions that govern the actual availability.

The impact of exponential demand on the life index

The failure to recognise the proved exponential growth in demand for resources introduces large errors into the so-called static reserve life index. Behrens (1973) has shown, with an assumption of growth rate of 3 per cent per year, the resultant reduction of the static reserve life index for some minerals (Table 1).

TABLE 1
Comparison of life indices for some minerals for the world (after Behrens, 1973).

Mineral	Static Index	Exponential Index
Iron Ore	400	86
Manganese	180	62
Aluminium	175	60
Copper	40	26

The calculation is done using the formula

$$e = \frac{\ln (r \times s + 1)}{r}$$

where, e = exponential reserve life index,
s = static reserve life index, and
r = annual growth rate.

This table shows that the static reserve life index is an inadequate measure of resource availability if demand is increasing exponentially. The estimate of 3 per cent growth rate is not critical. Fig. 2 (Behrens 1973) illustrates the relationship between the static and exponential indices at various rates of growth.

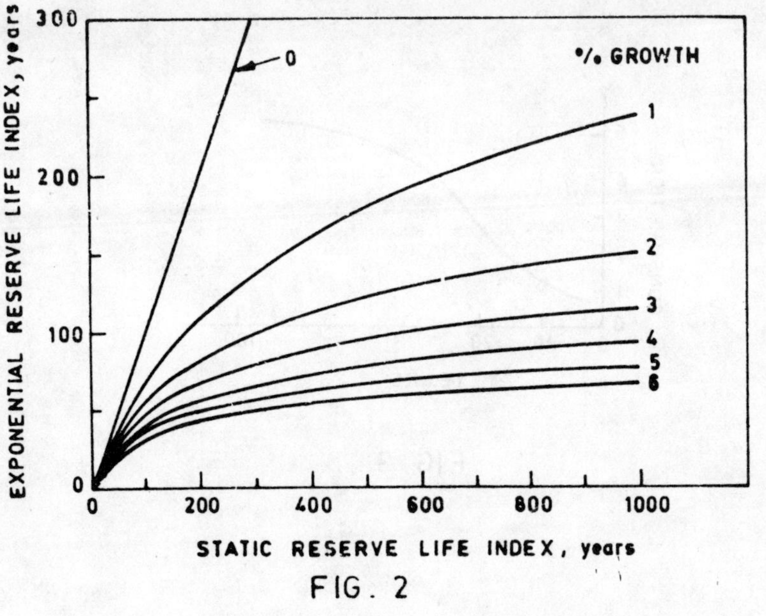

FIG. 2

This figure also points out another implication of exponentially growing usage rates. New discoveries increase the exponential life index much less than one tends to believe.

It is widely known that the developed countries particularly the USA, Western Europe and Japan, consumed the bulk of world mineral resources in seventies. Their share is particularly high for non-ferrous metals and energy. Leontief has shown that in the projections their high share continues but will decrease substantially over the period 1970-2000 as developing countries grow in relative economic importance. In general, the demand for mineral resources is roughly proportional to GDP but resource consumption will tend to increase with rise in income in the developing regions. Table 2 gives the rates of growth of GDP and the minerals.

TABLE 2

Growth rates of consumption of minerals
1970-2000 (Percentage) (After Leontief)

	Developed market	Developed Centrally Planned	Developing
GDO growth rate	3.6	5.1	7.2
Resources			
Copper	4.2	5.8	8.7
Bauxite	3.9	5.8	7.3
Nickel	3.6	5.4	7.1
Zinc	3.7	5.4	7.6
Lead	4.1	5.9	8.6
Iron	4.0	5.4	7.9
Coal	5.0	5.1	6.4

Walker (1979) has also advocated that rapid growth occurs in the early rather than the late stages of the process of industrialization which can be represented as shown in Fig. 3.

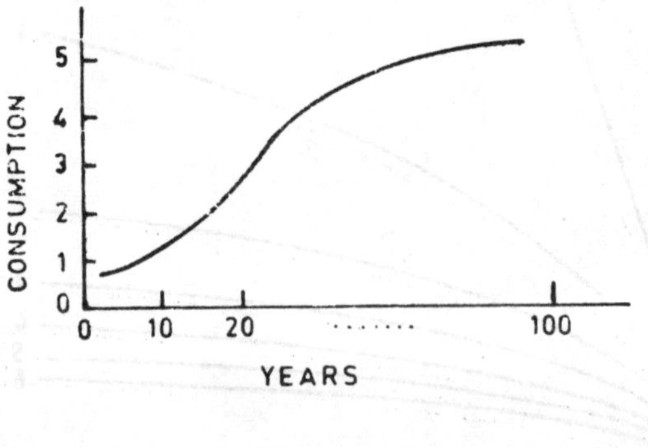

FIG. 3

The impact of other factors on the life index

There are few assumptions for calculation of static life index namely consumption will remain constant, grade will not decrease, no new discovery will take place, no technological change etc. These assumptions make the concept of life index highly conservative. Ores of lower grade than those mined at present will be mined in future with depletion of resource prices will increase making lower grade ore mining profitable. This fact will partially counteract the effect of exponential demand on static index.

There are some other factors also which will compensate partially for exponential growth in demand with falling grade prices which might put a restriction in demand and spur technological change. One of the area of technological change is in the technology of extraction, processing and distribution. Other important area is in the technology governing substitution. The notion that most materials can be substituted in various ways has attracted support. But there is limit to substitution specially for metal to metal. Recycling technology is another major area which helps in a big way the resource availability.

Realistically, no mineral resource will be completely exhausted. Rather the resources will come to exist only in such remote and dilute forms that they cannot be economically extracted. The total of all concentration of a given element in the upper crust which may eventually become exploitable form only a small fraction of the total amount of the element in the earths crust. This amount is measurable by the average crustal abundance as advocated by Engelhardt (1975) and depicted in Fig. 4.

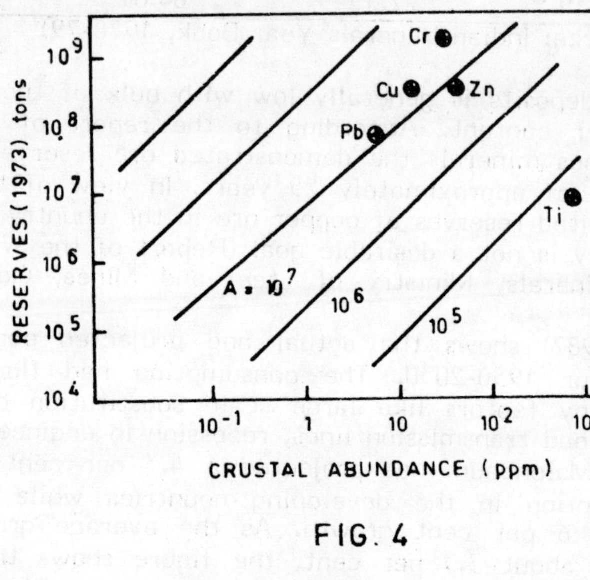

FIG. 4

COPPER IN INDIA

Copper is one of the important strategic non-ferrous metals essential for industrial growth and demand for copper has been growing over the years. The primary use of copper in India is in the electrical industry and the break-up of consumption is shown in Table 3.

TABLE 3

Percentage break-up of copper consumption in India

Winding wire industry	36%
Electrical goods, motors, transformers	6%
Cable and wires	14%
Automobile and auto-ancillaries, alloys	22%
Mint, ordinance, railways, etc.	12%
Others	10%

(Source: Rao, M.S., Scenario of Development of Copper, Seminar on Copper Industry in India, Khetrinagar, 1981).

India's reserves of copper are small which is summarised in Table 4. This table shows only the major reserves and respective States where they occur.

TABLE 4

Reserves of Copper in India

State	Measured	Indicated	Inferred	Total
All India	100.30	213.66	142.26	456.23
Bihar	82.15	90.99	44.93	218.56
M.P.		76.19	23.58	99.77
Rajasthan	16.19	22.19	64.67	103.05

(Source: Indian Minerals Year Book, 1978-79)

The grade of the deposits is generally low with bulk of them below 1.5 per cent in copper content. According to the report of the Working Group of Non-ferrous minerals the demonstrated ore reserves are expected to have a life of approximately 25 years. In view of the strategic importance and limited reserves of copper ore in the country it is advised that self sufficiency is not a desirable goal (Report of the Working Group on Non-ferrous Minerals, Ministry of Steel and Mines, Govt of India, 1980).

Fig. 5 (Shyam, 1982) shows the actual and projected consumption of copper in India for 1950-2030. The consumption had fluctuated very widely due to many factors like large scale substitution of copper by aluminium in overhead transmission lines, recession in engineering industry in mid-seventies. Malenbaum has projected a 4.3 per cent growth rate in copper consumption in the developing countries while the working group predicted a 6 per cent growth. As the average growth rate has been found to be about 3.3 per cent, the figure shows the projection for a growth rate of 3%.

Fig. 6 (Shyam, 1982) shows the consumption of copper super-imposed on the reserves of copper in India. Assuming in-situ reserve of about 450 million tonnes and average grade of 1.2 per cent copper content and 60 per cent of in-situ reserves are recoverable the metal reserve would amount to about 3 million tonnes. If domestic demands are to be met completely from these reserves it would run out in about 20 years from now.

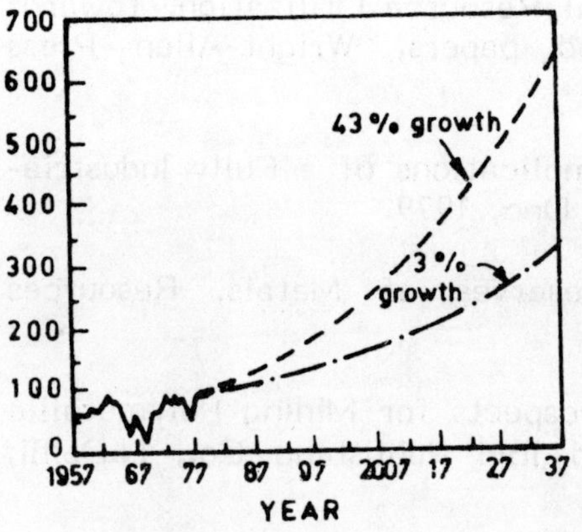

FIG. 5

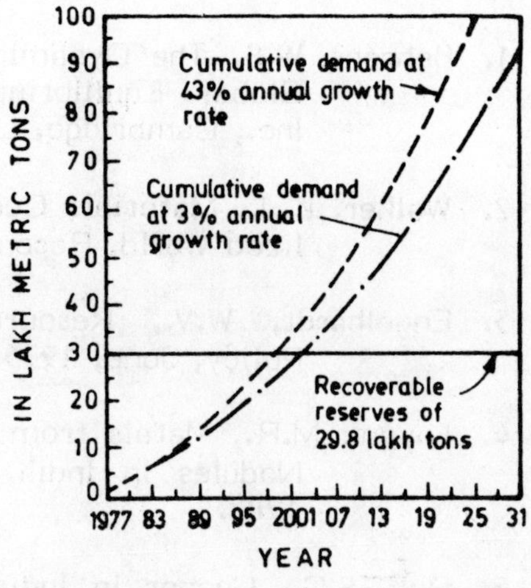

FIG. 6

There is also an important source of industrial feedstock in the form of scraps or secondary materials. Walker has reported that recycled copper including new scrap amounts to about 40% of current consumption globally. Even if the secondary supply meets 20-30% of our demand, then the life of land-based resources would be extended greatly. Also we must consider the supply of copper from mining of manganese nodules which is a reality now and we might get the supply of copper from this source in another 10 years. All these potential sources would extend the life index of copper ore in India substantially.

In an earlier paper, Ray (1984) had shown how logistic curve can be used to study the long term production capability under the dynamic concept of reserves and resources. This approach was based on the assumption that production of an exhaustible mineral commodity begins at zero, increases exponentially undergoing a period of more or less continuous growth. However, this exponential growth is a temporary phenomenon and a culmination of growth is reached in due course. Thereafter, there will be a decline in supply. This concept is also supported by Petersen and Maxwell.

CONCLUSIONS

The reserve life index is a poor measure of resource availability. There is no better index at present possibly due to lack of structure required to interrelate different informations. The life index calculated by some available mathematical formula and use of logistic curve will give some idea about the resource availability. But a system dynamics study using simulation technique will possibly provide a better understanding required for long term policies incorporating all the main factors influencing the supply and demand.

REFERENCES

1. Behrens, W.S., The Dynamics of Natural Resource Utilization, Towards Global Equilibrium. Collected papers, Wright-Allen Press Inc., Cambridge, USA, 1973.

2. Walker, K.J., Materials Consumption Implications of a Fully Industrialized World, Resources Policy, Dec, 1979.

3. Engelhardt, W.V., Resources and Reserves of Metals, Resources Policy, June, 1975.

4. Shyam, M.R., Metals from Seabed : Prospects for Mining Polymetallic Nodules in India, Oxford and IBH Publishing Co., N.Delhi, 1982.

5. Ray, S.C., Copper in India - Present and Future Outlook, Resources Policy, June, 1984.

ROLE OF SMALL SCALE MINING

C.S. Jha
Bihar State Mineral Development Corporation
Ranchi, India.

ABSTRACT

The socio-economic role of small scale mining in a developing country is important. India has vast reserves of mineral in small deposits which can be suitably worked for small scale mining. It promotes employment without requiring high skills or heavy investment. In the process, remote and backward areas can be developed and rural industrialisation can get a boost. To make a success of small scale mining the Central and the State Governments, as also the professional bodies, have to join hands for the purposes of amendment to legislations, provision of financial incentives, fixing up agencies and means for exploration, project planning and implementation, marketing and building of infrastructure. Various aspects governing the operations such as legislative, social, economic, financial, technical and environmental, will have to be kept in view and examined for appropriate action at the level of the industrialists, the State Government and the Central Government. Educational institutions, State Mineral Corporations and the technical bodies can render valuable assistance. There is a need for periodical review by a high level set up for suggesting corrective steps from time to time.

INTRODUCTION

Small scale mines are widely prevalent throughout the world. About 16% of the world production of non-fuel minerals in terms of value comes from small scale mines. In some countries it is much higher, running upto 100%. In case of a number of minerals small scale mining provides the only way for mining because of the very nature of occurrence of the mineral deposits. There is also a distinct category of small mines which cover small parts of large mineral deposits where an initial small scale mining develops into large scale. The paper surveys the status of small scale mining and suggests measures for promoting small scale mining.

SMALL SCALE MINING WORLDWIDE

Table 1 in the appendix gives a list of developing countries with significant small scale mining activities. This list is indicative, and not exhaustive. Bolivia has 3000-4500 small mines, Brazil about 4000, Peru 2500-3000 and Mexico about 2500. The 380-400 small mines of Zimbabwe employ 8000-9000 workers. Small mines are reported to account for about 10% of China's coal output and about 7.3% of its iron ore output. Thailand has more than 1000 small scale mines. Malaysia and Indonesia about 500 each and Burma about 200. Even in the USA more than 50% of the production of coal is coming from small scale mines, the scale being adjudged by their standard.

The National Resources and Energy Division of the United Nations Department of Technical Co-operation for Development organised a study tour of coal specialists from developing countries to visit various small scale coal mines in China. There has been remarkable success in developing small scale coal mines in China in an effort to exploit indigenous energy resources for regional economic development. In 1982, out of the country's overall coal production of 666.32 million tonnes, local small mines produced 316.4 million tonnes (47.5% of the country's total) from 20,000 mines and employed 802,000 persons, not all of them are full-time workers at the mines. The local mines' contribution of 47.5% came from provincial (10.6%), prefectural (6.4%) county (8.6%) and commune (21.9%). The capacity of local mines ranged from several thousand tons to a million tons a year. "Given the physical limitations of transport and other infrastructure, the vast requirements of industrial centres could not have been met merely by depending upon the country's major coal fields. It was in this context that emphasis was laid on the development and exploitation of locally available resources. The local mine itself is generally simple, with unpretentious buildings serving the needs of administration and workshop. Unlike the large mining complexes, these small units do not provide housing to the employees, as they are drawn from nearby rural settlements".

In India, about 1900 non-coal small mines submit returns under the Mines Act because either they have underground workings or the depth of open workings exceed 6 metres or explosives are used. Thousands of additional mines which are known to be producing industrial minerals like clay, sand, stone aggregates, etc. either do not come under the umbrella of the Mines Act and therefore do not submit any returns or they simply do not keep the authorities informed about their activities, taking advantage of remoteness and isolation of such mines. Table 2 in the appendix gives the number of small non-coal mines in India according to employment.

SMALL SCALE MINING - DISTINCTIVE FEATURES

Small scale mining has an important socio-economic role to play. This is true particularly for the underdeveloped backward and rural areas. It helps in generating employment without requiring high skills or heavy investments. Such employment does not remove the workers from their villages and does not, therefore, add to the problems of overpopulation in urban areas. Most of the equipment required for such mines can be produced by local workshops. The activities are man-intensive and the requirement of energy is relatively low. In some of the remote areas, small scale mining can supplement seasonal employment in essential occupations like agriculture. The produce from the small scale mines can be useful for opening small mineral-based industries which in turn create further potential for employment. The problems of long distance transport of minerals are also minimised. It can thus be seen that by resorting to small scale mining at suitable sites there can be much greater utilisation of manpower for the creation of national wealth, thereby improving the living standards.

BACKGROUND AND DEFINITIONS

In the past seven years, there have been a number of important occasions when the mining community has discussed small scale mining. The first International Conference on Small Scale Mining was held in Mexico in 1979. Subsequently, a Workshop on Mineral Policy for Small Scale Mining was held at New Delhi in 1984. It just preceded the 12th World Mining Congress. The Congress included a round-table discussion on Small Scale Mining. Under the aegies of The Mining, Geological & Metallurgical Institute of India several regional seminars discussed the subject of Small Scale Mining in India and finally a National Seminar on Small Mineral Deposits - Their Development and Industrial Possibilities was organised in 1985. The conclusions and recommendations have been made for the consideration of the Government of India.

Definition of Small Scale Mines

Opinions differ a great deal over the definition of the Small Scale Mines. The following definition has been recommended:

"A small scale mine is a mine which produces a mineral in the raw unprocessed form, as measured at the entrance to the mine, upto the limits given in the Table 3 annexed, subject to the maximum investment in plants, machinery and equipment of Rs. 6.0 million without any mechanised beneficiation and processing unit and of Rs. 10.0 million with such a unit".

Definition of Small Mineral Deposit

"A small mineral deposit is one whose total geological reserve in the possible category is either less than or equal to the amount determined by multiplying the figure of maximum defined yearly production level of small scale mine, exploiting the mineral, by a factor of 120 (20 for the normal life of a mine x 1.5 against the average limit of recovery of mineable reserve being 2/3rd x 2 for meeting the future need of expansion x 2 for converting mineable reserve into geological reserve. Thus the upper limit of geological reserve of a small deposit is linked with the definition and need of the small scale mine. It is 120 times the yearly maximum production of the mine. Although the actual reserves in many small deposits are much smaller than the limits given in the table, the definition does not exclude them from the category of small deposit and the higher limit is to allow the small scale mine operating the deposit to be financially viable (reserve limit being on pit mouth value basis). It is, therefore, recommended that the above definitions be accepted and adopted immediately in order to make a beginning and a designated body may be asked to revise it after giving effect to it for a period of 5 years or so, the revision being made preferably before a Five Year Plan begins".

NATIONAL MINERAL POLICY AND SMALL SCALE MINING

Whatever is done in respect of small scale mining has to be in conformity with the National Mineral Policy. The objective of the National Mineral Policy is to regulate the development of national mineral resources to subserve the broad national objective of rapid economic growth, national security, generation of employment potential and removal of regional imbalance. The present mineral policy of India is based on the second Industrial Policy Resolution of 1956 under which the minerals have been divided into three categories for their regulation and control.

1st: Coal, lignite, mineral oils, iron ore, manganese ore, chrome ore, gypsum, sulphar, gold and diamond, mining and processing of copper, lead, zinc, tin, molybdenum, wolfram and atomic minerals.

2nd: All minerals other than minor minerals.

3rd: Minor minerals.

While the minerals of the first category are predominantly in the State sector, minerals of the second category are expected to be progressively State owned and the minerals of third category are left for the initiative and enterprises of the private sector.

The Mines and Minerals (Regulation and Development) Act 1957 and the Mineral Concessions Rules 1960 are the instruments through which the Government is implementing the provisions of the Industrial Policy Resolution regarding minerals. The State Governments are authorised under the rules to grant leases directly for major minerals other than those given in the first schedule in the Mines and Minerals (Regulation and Development) Act for which prior approval of the Central Government is necessary. For minor minerals, each State Government frames its own rule and it has the exclusive right for grant of leases.

It would thus be seen that the Government of India so far has been following a definite and conscious mineral policy with pragmatic and nationalistic outlook which is in line with the general world trend. But all this has been done and endeavoured to be done mostly through large scale mining sector. What is now necessary, as will be emphasised in this paper, is a conscious, clear and codified National Policy on development of small deposits and small scale mining for achievement of some positive objectives.

The Seven Five Year Plan has specifically drawn attention to the fact that the life of reserves of certain minerals such as copper, coal, high grade iron ore, magnetite, kyanite, lead, chrome, magnesite, manganese ore, high grade rock phosphate, massive sillimanite and zinc is less than 50 years. Life indices of known mineral reserves at the projected rates of depletion is given in the Table 3 annexed.

While massive efforts have to be made for the expansion of mineral deposits and scientifically planned management of mineral resources, the small deposits and small scale mining, wherever possible, cannot be ignored.

DEVELOPMENT OF SMALL SCALE MINING - SOME RECOMMENDATIONS

We can consider various aspects of small scale mining under the following broad heads:

Legislative aspects

A large number of statutes are applicable in case of mines and almost all of them are applicable to small scale mines also. Many of the provision, however, are unnecessary and too rigorous to be implemented by the small scale mines. Two main problems concerning legislative aspects are (a) rigorous provisions and (b) impracticability of application. It is, therefore, necessary to repeal or modify provisions of the law so as to encourage small scale mining operations. It is suggested, therefore, that a technical committee should be constituted by Govt of India to make detailed study of the problems and spell specific recommendations. Under no circumstances, however, the provisions for safety should be ignored.

S o cial aspects

For small scale mining, the social aspects are of greater importance than technical ones. The health and willingness of work persons have to be ensured. Because of inaccessibility during rains, many mineral deposits are worked on seasonal basis. For such mines relaxations in the statutory obligations regarding continuous employment have to be given. Such relaxations may be given in a suitable manner to allow the landless workers to devote themselves to agriculture during rains.

Employment of local people has to be encouraged, particularly while developing small deposits under tribal and backward areas. The problem of acquisition of land has social ramifications. This has to be tackled in a judicious way.

The work force requires some amount of training. The State Government through Training Institutions and concerned directorates should organise training-cum-orientation programmes which are different from technical training courses. It seems desirable to make elaborate study of socio-economic effect of small scale mining. This can be done by a competent organisation under Government patronage.

Economic aspects

To make a success of small scale mining, economic aspects have to be kept in mind, and for this purpose, marketability has to be assured. Adequate market intelligence and marketing facilities should be provided and this can be done by State through State Mineral Development Corporations which can have special cell for the purpose. The large consumer industries like the Steel Plants, Refractories, Cement Plants, etc have to be asked to procure part of their requirements from the local small producers. Many of the minerals extracted by small scale mining have good export potential, and hence they have to be helped by appropriate fiscal and financial policies to increase earning of foreign exchange.

The benefits of abolition of export duty and tax relief as applicable to other export materials should be extended to minerals sector also. Wherever possible, equipment can be given on loan from the State Mineral Development Corporations.

Financial aspects

Small scale mining sector must have a sound and favourable financial arrangements for operations such as mining, exploration, processing, transportation and marketing. Fiscal and financial incentives applicable to small scale industries should, therefore, be extended to mining industry also.

After preliminary prospecting, detailed exploration for commercial operation should be taken up either by the State Directorate of Geology or by the State Mineral Corporation for which a special mineral development fund is to be created for processing and beneficiation. The leasees should bear part of the cost and the balance should be paid by the State Government. Lessees should be provided easy term loans at low interest by the State Financial Corporations and 10% of the cess should be devoted for the mineral development fund to which the Central Government should also suitably contribute.

Technical aspects

The basic need is the preparation of state-wise folio, maps of different small mineral deposits and this can be done jointly by GSI, IBM, State Directorates of Geology and Mining and State Mineral Development Corporations. Indicative geology can be done by the State Directorate of Geology and the detailed geology can be organised jointly by the State Mineral Corporations and the lessee.

Development plans have to be prepared carefully and for this purpose feasibility report/project report can be made by an expert body under the State Corporation. So far as mining operations are concerned, these are likely to be mostly by opencast, the planning of which is comparatively easier. But mechanisation to some extent can be introduced when it becomes absolutely necessary depending on the situation. By opencast mines, however, the planning of benches, haul roads, dumping, drainage, etc will have to be given careful thought. Small scale mines may not be able to maintain their own fleet of equipment. It is suggested that equipment centres can be opened for serving a group of mines, the cost of which can be shared by the mines taking advantage of the centre.

Basic infrastructures such as supply of power, connecting roads, transport facilities, communication, medical facilities, etc would be needed. Construction of roads obviously should be taken up by the State Government and the mine owners can take responsibility for feeder roads from the main roads. Railways have to be planned by the State Government in consultation with the Central Government, the economics of which would depend on the total transport of minerals and other goods.

Technical training for manpower working in the mines would have to be organised and help can be sought from Government institutions, mining associations and local consultants. The big mines in the neighbouring areas should also extend helping hand to small mines. Technical assistance and consultancy services should come from organisations such as Indian Bureau of Mines, MGMI and others.

The future prospects of mineral-based industries is quite good. Feasibility reports for such industries have to be prepared and this should be done by the Mineral Development Corporation of the State in consultation with Industrial Development Corporation.

Environmental aspects

Because of opencast mining, land damage will occur which will create ecological problems. It is necessary to plan for reclamation of land followed by afforestation. This responsibility should rest with the owner of the mine. Where reclamation is not possible, proper water reservoirs should be developed. Due care has to be taken to deal with effluents. The planning itself will have to take into account the maintenance of healthy environment and ecological balance. For certain different problems regarding various Conservation Act, there should be a rapport between the Department of Mines and Geology and the Forest Department of the State and for all these purposes a State Mineral Advisory Board should be constituted which should be broad-based.

While granting a lease preference should be given to those who are likely to set up mineral-based industries and the mine owners of small scale mines should be liberally exempted from appointing full-time qualified mining engineers or geologists as per the existing rules.

CONCLUDING REMARKS

Emphasis and priority have to be accorded to small scale mining for recovering our mineral wealth, which will lead to rural industrialisation, development of backward areas and creation of employment - all making substantial contribution to the national economic well being. Geologists, mining engineers and planners must join hands to create the right atmosphere for its success.

TABLE 1

Developing Countries with significant
Small Scale Mining activity

Country	Minerals
Latin America	
Bolivia	Tin, antimony, lead, tungsten, gold
Brazil	Semi-precious stones, berylium, gold, titanium (ilmenite), chromium, diamonds, mica (vermiculite)
Chile	Copper, sulphur, gold
Mexico	Niobium (pyrochlore) mercury, fluorspar, tin, silver, opal
Peru	Lead, antimoney, tungsten, gold, zinc, silver
Africa	
Mozambique	Semi-precious stones, tantalum, mica
Nigeria	Tin, niobium-tantalum (columbite), gold
Sierra Leone	Diamonds
Zaire	Diamonds, tin
Zimbabwe	Chromium, gold, lithium, precious stones
Asia	
Burma	Tin, tungsten, antimony
China	Tin, antimony, tungsten, iron, coal
India	Limestone, iron, mica, manganese, chinaclay
Indonesia	Tin, gold
Malaysia	Tin, titanium (ilmenite)
Thailand	Tin, antimony, tungsten
Turkey	Chromium, antimony, mercury, lead, zinc, asbestos

Source: Werner Gocht: The Importance of Small Scale Mining in Developing Countries, Natural Resources and Development, Vol.16, Hannover,1982.

TABLE 2

Number of small non-coal mines* in India
- According to Employment** (1980***)

(a) Overall figures:

Total employment	Total number of Mines	Out of which, number of mines having underground employment of	
		Upto 75	76 to 150
Upto 50	1159	231	3
51-151	422		
151-250	102	3	9
151-400	63		
Total	1874	234	12

* Data relates to only such mines as submit 'returns' under the Mines Act.

** Average daily employment

*** Current figure will not be significantly different.

(b) For some main minerals:

Mineral	With overall employment of				Total
	Upto 50	51-151	151-250	250-400*	
Limestone	105	81	22	9	2 5 2
Iron Ore	69	66	26	10	1 9 9
Stone	139	40	3	3	1 8 6
Mica	143	27	6	3	1 7 6
Manganese Ore	90	41	6	9	1 6 2
Chinaclay	110	30	4	3	1 4 9
Steatite	94	10	2	3	1 0 9
Fireclay	62	14	2	1	9 8

* Some of these mines may be highly mechanised and would need to be excluded from the category of small mines. Their number however is not large.

TABLE 3

Group	Minerals	Production	20 years life with 2/3rd% ultimate recovery from mineable reserve	2 times for future expansion	Total mineable reserve in million tonnes	Total geological reserve possible category
Group I	Asbestos 'A' & 'B' quality	Upto 250 tonnes per year produce	250 x 15 x 20	x 2	0.015	0.03
Group II	Mica Ore(3% mica basis) Fluroite (85% CaF_2 basis)	Upto 1000 kg per year produce	1000 x 1.5 x 20	x 2	0.06	0.12
Group III	Gold Ore(5 ppm basis Sillimanite)	Upto 2500 tonnes per year produce	2500 x 1.5 x 20	x 2	0.15	0.30
Group IV	Kyanite,Baryte, Graphite (60%C. basis)	Upto 5000 tonnes per year produce	5000 x 1.5 x 20	x 2	0.30	0.60
Group V	Phosphorite/ Apatite : Manganese Ore (25% Mn). Chromite, Magnesite, Talc/ Soap-stone (powdery) Tungsten Ore (0.2% WO_3) Calcite, Chinaclay Fireclay	Upto 10000 tonnes per year produce	10000 x 1.5 x 20	x 2	0.60	1.20

TABLE 3 (Contd.)

Group	Minerals	Production	20 years life with 2/3rd% ultimate recovery from mineable reserve	2 times for future expansion	Total mineable reserve in million tonnes	Total geological reserve possible category
Group VI	Lead Zinc Ore, Copper Ore, Gypsum, Feldspar, Quartz, Silica Sand	Upto 25000 tonnes per year produce	25000 x 1.5 x 20	x 2	1.50	3.00
Group - VII (A)	Bauxite, Iron Ore, Limestone, Dolomite Stowing Sand, Stone Aggregate, Brick Earth, Ochre, other minor minerals	Upto 50000 tonnes per year produce	50000 x 1.5 x 20	x 2	3.00	6.00
(B)	Coal Lignite	- do -				

THE FUTURE OF LARGE, LOW GRADE,HARD-ROCK TIN DEPOSITS PERSPECTIVES AND PROBLEMS

Dr. Camillo Premoli
International Mineral Research
Sydney, Australia.

ABSTRACT

The price of tin is at historically low levels and tin is in oversupply. Present production is largely from alluvial deposits worked mainly by small and medium-sized producers. These producers, often undercaptilized and working deposits of declining grade, are faced with rising production costs. It is expected that many will cease production in the future. Interest is therefore beginning to focus on the viability of exploiting large hard-rock tin deposits amenable to low-cost bulk mining methods.

Among the countries which appear to have very large reserves of hard-rock tin deposits is China, and it is anticipated that in the future China will be a major supplier of tin. This paper originally appeared in Natural Resources Forum (Vol.9,No.2,1985) and is being reported with permission.

INTRODUCTION

Tin is a metal which has unquestionably reached maturity in its industrial application and, in fact, may well have entered into a phase of decline. Figure 1 shows the poor performance of tin relative to other metals during the period 1970-1983. For metals other than tin, consumption increased between 11.4% and 51.5%, whereas tin consumption declined 7.5%. Without a dramatic upswing in the consumption of tin, major changes in the pattern of supply are inevitable.

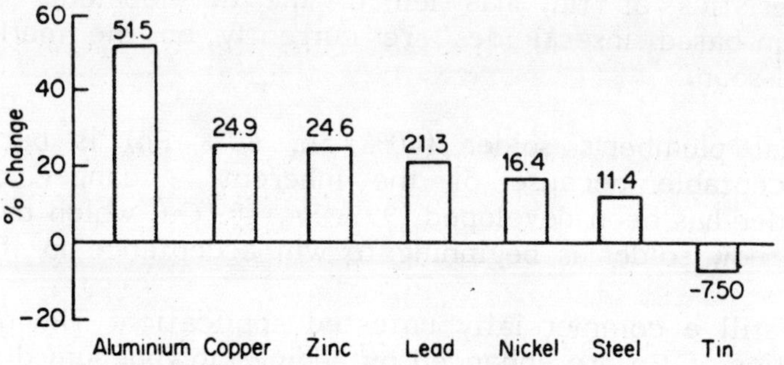

Fig 1 — Worldwide changes in consumption of selected metals, 1970-1983.

The tin market continues to be in structural oversupply, and with the uncertainty surrounding the future applications of tin the permanent closure of high-cost mines would appear likely. It must be stressed, however, that the potential availability of tin metal, even with such closures, would still be very large.

Table 1 lists the world's stanniferous areas and Fiture 2 shows world tin production and consumption by the three main economic groups.

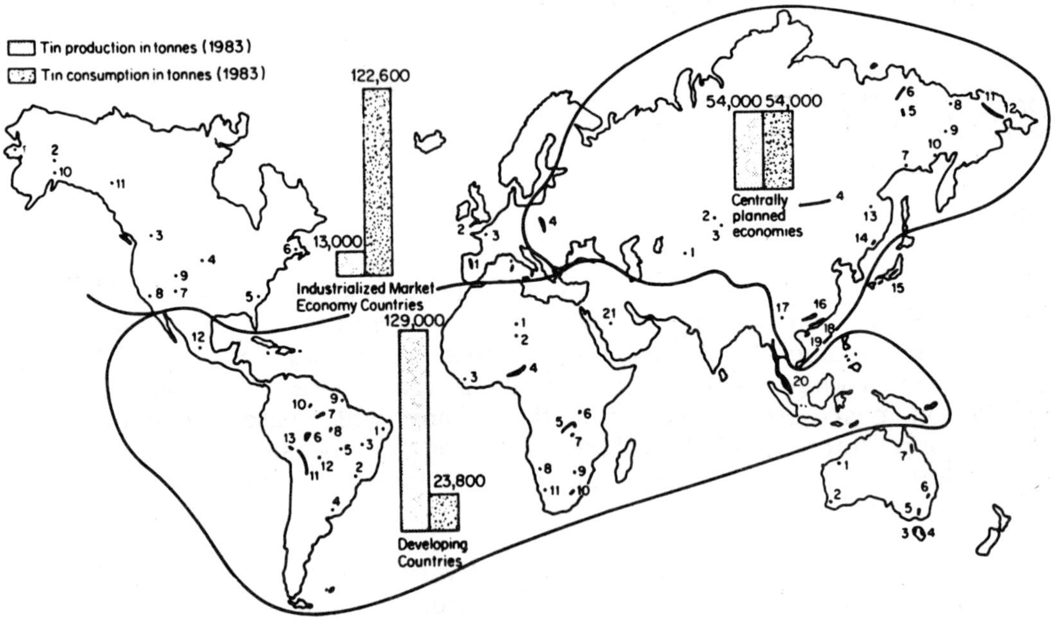

Fig 2— World stanniferous areas, tin production and consumption by economic group. The names of the localities are listed in Table 1. Source: US Bureau of Mines.

The best hope for the future of tin, at least in the long term, seems to lie not in the development of new geographical markets, but rather in the discovery of new uses. The International Tin Research Institute (ITRI), based in London and funded by some of the major tin-producing countries, is devoting itself to this objective. Three main new uses for tin have been stressed.

Tin chemicals. Tin is one of the comparatively few metals with no harmful effects on the environment or the human body. This fact, and the unique chemical characteristics of tin, has led to the development of new biocides. Three tin-based insecticides are currently on the market and more are expected soon.

Solders. Traditional plumber's solder (40% Sn, 60% Pb) is becoming increasingly unacceptable because of the inherent danger of lead. An alternative solder has been developed (99% Sn, 1% Cu) which overcomes the problem. This new solder is beginning to win acceptance in the USA.

Bearings. This is still a commercially untested application. The naturally lubricating properties of tin are enhanced by adding molybdenum disulphide, thereby permitting the use of tin in heavy-duty bearing applications.

Tin producers can expect even more diversified uses for tin in the future. Work even now is in progress on tin-zinc alloys to replace toxic cadmium plating and on the use of tin chemicals in coal-liquefication processes.

In a situation where tin has been considered merely as a "mature" commodity with a settled future it may now be entering into a new phase as a "high technology" metal. This could mean increased demand for tin in the long term.

TABLE 1 – Names of world's stanniferous areas (see Fig 2 for locations)

NORTH AMERICA
1. Seward Peninsula, Alaska, USA
2. Manley Hot Springs, Alaska, USA
3. Sullivan Mine, BC, Canada
4. Black Hills area, S Dakota, USA
5. Southern Appalachian area, USA
6. East Kemptville, Canada
7. Black Range area, New Mexico, USA
8. Gorman area, California, USA
9. Climax mine, Colorado, USA
10. Coal Creek, Alaska, USA
11. Seagull Batholit JC Sharn, Yukon, Canada
12. Durango, Mexico

SOUTH AMERICA
1. Seridozinho, Brazil
2. São Joao del Rey, Brazil
3. Paramirin-Rio de Cuentas, Brazil
4. Encruzilhada do Sul, Brazil
5. Central Goias, Brazil
6. Rondonia, Brazil
7. Rio Piting-Mapuero, Brazil
8. Rio Xingu, Brazil
9. Amapa, Brazil
10. Sucurus, Brazil
11. Bolivian tin belt, Bolivia
12. Ascencion de Guarayos, East Bolivia
13. San Refael-Quenamari, Peru

AFRICA
1. Haggar area, Algeria
2. Air area, Niger
3. Tawalpa area, Sierra Leone
4. Jos Plateau, Nigeria
5. Maniema tin belt, Zaire
6. Central Africa tin belt, Rwanda-Burundi
7. Manono-Kitolo, Katanga, Zaire
8. Brandeberg tin field, Namibia
9. Bikita-Kamativi tin field, Zimbabwe
10. Bushveld tin field, South Africa
11. Orange River W-Sn pegmatites, RSA

EUROPE
1. Portugal and Spain
2. Cornwall, England
3. Erzgebirge, GDR-Czechos
4. Allier, France

ASIA
1. Altai Mountains, USSR
2. Naryn, Kazakstan area, USSR
3. Kalba, East Kazakstan, USSR
4. Transbaikal (Zabaykal'ye) area, USSR
5. Yana-Adycha area, USSR
6. Ege-Khaya-Deputatskiy area, USSR
7. Ust'-Omchug, Batugychag area, USSR
8. Bilibina area, USSR
9. Kolyma area, USSR
10. Galimyy area, USSR
11. Krasnoarmeyskoye area, USSR
12. Iul'tin area, USSR
13. Khingansk area, USSR
14. Lifduzin-Khrustal'nyy area, USSR
15. Honshu Island, Japan
16. Kwagnsi-Dachang area, China
17. Kochiu area, China
18. Coastal zone, China
19. Hainan Island, China
20. Malay Peninsula
21. Jabal as Sisilah, Saudi Arabia

AUSTRALIA
1. Pilbara, Western Australia
2. Greenbushes, Western Australia
3. North West Tasmania
4. North East Tasmania
5. Albury-Ardlethan, NSW
6. New England District, NSW
7. Herberton-Cooktown area, Qld

The enormous imbalance between the industrial and developing world in terms of tin production and consumption is striking. By contrast, the centrally planned economies seem to be in a more balanced position.

From the production point of view, interest is beginning to focus on the viability of exploiting large hard-rock tin deposits amenable to low-cost bulk mining methods. During the last few years Brazil, Canada, Australia, obsolete manufacturing techniques entirely and opt for off-the-shelf advanced techniques already tested and adopted in the industrialised countries, at least for the most vital part of their consumer sector.

The case of two leading developing nations is very telling: Brazil, in spite of its rapid industrialisation, has seen a decline in domestic tin consumption and this, coupled with its vigorous tin production, has increase tin available for export from 3000 t in 1982 to 9200 t in 1983 (Mining Annual Review, 1984). China, which is also bent on an ambitious program of expansion of its industrial base, is trying to limit its tinconsumption. At a national conference held at Gejiu early in May 1982 it was considered that excessive domestic consumption was one of the three main problems of the Chinese tin industry (Mineral Yearbook, 1982). It is known that in the last few years Chinese domestic tin production has, in fact, been reduced.

It may be contended that the lower tin prices postulated by the weakening of the ITA in the introductory paragraph could also stimulate consumption. However, prices a re already historically low, and no such increase has been noted. As for new uses of tin, it is unlikely that these will have a significant impact on the tin market before the end of this decade and, in the meantime, fundamental changes in the market structure will have taken place.

The most realistic projections for tin demand for the year 2000 are between 160 000 and 200 000 t of tin for the market economy countries. In the present study the figure of a tin demand of 186 000 t for the year 2000 is considered the most acceptable on the basis of the present knowledge of the market.

Figure 3 shows the consumption of tin from 1965 to 1985. Before 1981 the tin market was characterised by a very elastic supply and steadily increasing demand. The mid 1980s is a period of depressed demand and depressed tin prices. This period is expected to be followed by one of inherently inelastic supply and steady demand, rising gradually to 186000 t/year by the year 2000.

THE PRESENT SITUATION

Large hard-rock tin deposits are increasingly being exploited in Australia, Czechoslovakia, China, most recently in Canada (East Kemptville) and possibly in France (Allier) and in Brazil (Potosi Hill and Central Goias).

In the event of a collapse of the International Tin Agreement and the elimination of the protection that its pricing structure affords, the well-capitalised and technically efficient miners will have a distinct advantage over the alluvial miners, many of whom are under-capitalised and using outdated equipment.

MINING CONSIDERATIONS

In discussing the impact of large primary tindeposits on the tin industry the first problem is to define realistically the minimum grade and tonnage required to support a large tin-mining operation at foreseeable tin prices. Figure 4 gives the tonnages and grades of several major hard-rock tin deposits based on actual figures which have been released. Note that China has very large reserves of very good grades of tin ore. China has the potential to be a major producer of tin in the future.

Some authors have recently proposed a target scale for a profitable tin mine in the order of 40-60 mt of ore with a minimum head grade of 0.3-0.4% Sn and a projected recovery of at least 60%. These figures, of course, are somewhat viable, depending upon beneficiation and mining costs, and the available infrastructure.

Geologically, the type of primary tin deposits that can come within the grade-tonnages parameter indicated above and be amenable to low cost bulk mining, either by underground or open-cast methods, are:

1. subvolcanic hydrothermal deposits of stock work or referred to as "porphyry tin");
2. greisen-type deposits in all its numerous subtypes;
3. large limestone-replacement deposits; and
4. large pegmatitic deposits.

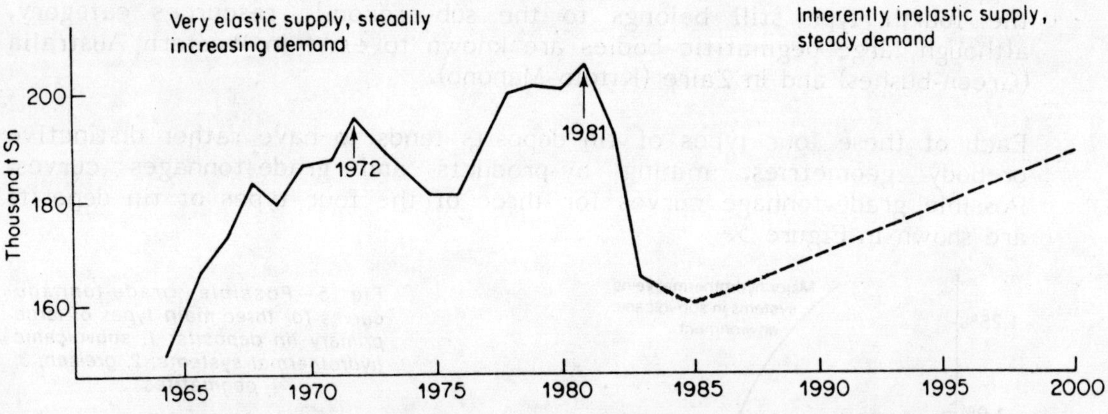

Fig 3—Past and projected tin production in market economy countries. The shaded area represents a period of probable phasing-out of high-cost producers.

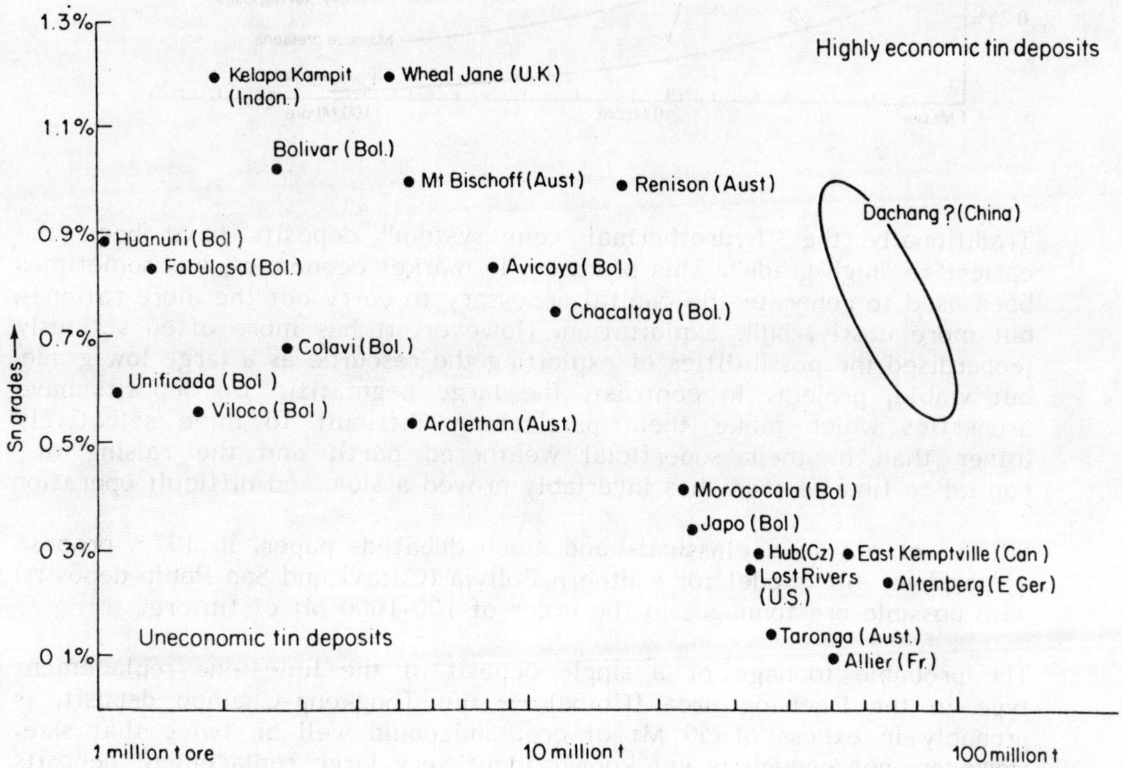

Fig 4 – Tonnage and grade of some major primary tin deposits of the world. The clustering of Bolivian deposits to the left of the figure indicates a serious ore reserve problem. The data for Dachang (China) are estimates by Western experts.

Only the first three types are economic and, therefore, mining realities; the fourth type still belongs to the sub-economic resources category, although large pegmatitic bodies are known to exist in Western Australia (Green-bushes) and in Zaire (Kitolo-Manono).

Each of these four types of tin deposits tends to have rather distinctive orebody geometries, mining by-products and grade-tonnages curves. Possible grade-tonnage curves for three of the four types of tin deposits are shown in Figure 5.

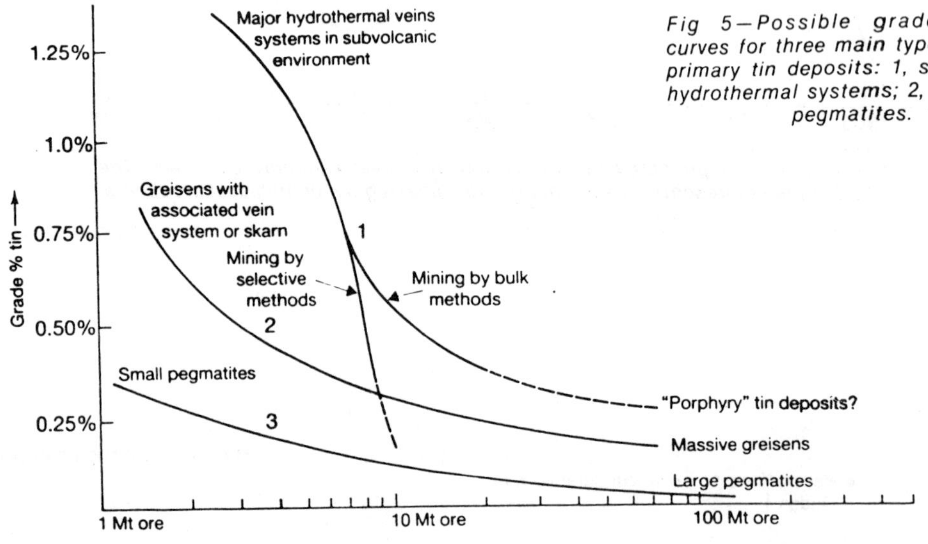

Fig 5—Possible grade-tonnage curves for three main types of large primary tin deposits: 1, subvolcanic hydrothermal systems; 2, greisen; 3, pegmatites.

Traditionally the "hydrothermal vein system" deposits have been the easiest to "high-grade". This practice, in market economies, has sometimes been used to generate the capital necessary to carry out the more rational, but more costly, bulk exploitation. However, it has more often seriously jeopardised the possibilities of exploiting the resource as a large low-grade, but viable, project. In contrast, the large pegmatitic tin deposits have properties which make them particularly difficult to mine selectively (other than in their superficial weathered part), and the raising of capital to finance them has invariably proved a slow and difficult operation.

Sillitoe et al., in a classical- and much debated- paper, in 1975, propose a "porphyry tin" model for southern Bolivia (Catavi and San Pablo deposits) with possible ore tonnages in the order of 100-1000 Mt of tin ore.

The probable tonnage of a single deposit of the limestone replacement type in the Dachang area (China), the Tongkeng-Champo deposit, is probably in excess of 50 Mt of ore, and could well be twice that size. However, not enough is yet known about very large replacement deposits of the Dachang (China) type to generalise their grade-tonnage curves.

From the mining point of view, large primary tin deposits often display characteristic (rock stability, orebody geometry, etc) which favour low cost mining methods. In Bolivia, Catavi is presently mined by block caving methods, and so is Altenberg in the German Democratic Republic. The geometry of the Tongkeng-Champ (Dachang) tin deposit in China seems very amenable to modern, underground trackless methods and a throughput in the order of 10 000 t/d of ore could easily be achieved when the mine is fully developed. The Renison tin mine in Australia has, for a number of years been using underground trackless methods and achieving very low mining costs. As a consequence, before the imposition of the ITA tin quotas, Renison was the most profitable mine in Australia.

IMPACT OF NEW DEVELOPMENTS

The large ore tonnages of the "new generation" hard-rock tin deposits permit a direct comparison with porphyry copper deposits, and their world-wide implications in terms of supply are worth examining.

The average mine life of new large tin-mines brought to production can be calculated using Taylor's rule:

minelife (in years) = 6.54 x $\sqrt[4]{\text{reserve tonnages}}$ (in millions of tonnes).

This rule offers a good empirical relationship between reserve tonnages and mine life, and can be directly related to production rate. Above all, the rule is simple and appears to work well for most types of deposits, whether they are mined by underground or surface methods. It is also quite adaptable to comparatively new types of deposits in untested conditions, and has proved remarkably accurate for projects in developing countries. This rule is not of course, related to the financial aspects of a project, and does not take account of variation in grade, development and infrastructure costs or production costs, taxation, etc.

Table 3 lists known large primary tin deposits by location, type, resource tonnage and ore-grade.

Some of these deposits are already being mined while others are under development and will be producing shortly. Still others are in the feasibility study stage.

Using Taylor's rule to calculate the mine life for selected mine tonnages, the possible impact of the development of large primary tin deposits on the tin supply for the market economy countries has been calculated and shown in Table 4. Combining the data in Tables 3 and 4 suggests that kthere is the potential for the production of large quantities of tin from primary tin deposits.

At present the pattern for the tin industry is one of domination bysmall and medium-sized producers, although there are kabundant indications that this pattern will be irreversibly changed by the extended period of depressed tin prices currently being witnessed. Low tin prices have already forced some tin producers out, and more are expected to follow. As the process continues, conditions favouring the development of large hard-rock tin deposits can be expected.

The geological favourability for large primary tin deposits seems more evenlyspread worldwide than for alluvial deposits (largelydependent for their formation on tropical conditions), and therefore new and significant tin suppliers are likely to emerge. Countries with good hardrock tin potential, adequate development capital and acc ess to advanced technology will become significant tin producers before the end of this country. With the notable exception of Brazil it is likely that tin mining

TABLE 2 – Tin production of the market economy countries (in tonnes)

	1972	1976	1980	1981	1982	1983
Members of the 6th ITA						
Australia	11 997	10 389	11 588	12 925	12 615	9578
Indonesia	21 766	23 418	32 527	35 268	33 800	26 554
Malaysia	76 830	63 401	61 404	59 938	52 342	41 364
Nigeria	6731	3710	2527	2383	1708	1450
Thailand	22 072	20 453	33 685	31 474	26 207	20 048
Zaire	5960	3950	3159	2346	2174	2004
Total	145 356	125 321	144 890	144 334	128 846	100 998
Non-members						
Bolivia	32 405	30 315	27 271	29 830	26 773	25 036
Brazil	2763	5482	6930	8297	8279	13 083
UK	3327	3323	3028	3870	4175	4069
Total	50 944	54 679	55 510	60 366	61 754	71 402
Total	196 300	180 000	200 400	204 700	190 600	172 400

TABLE 3 – Large primary tin deposits

Name	Locality	Type	Resource tonnage (Mt)	Grades (% Sn)	Comments
Allier	France	Greisen (?)	50	0.13	Under investigation by Penarroya
Altberg	GDR	Greisen	60–80	0.2–0.3	Underground mining by block caving
Catavi	Bolivia	Subvolcanic vein system, "porphyry tin"?	42 100–1000	0.22 0.2–0.5	COMIBOL (1982) Sillitoe *et al.* (1975)
Chorloque	Bolivia	Breccia type, "porphyry tin"?	Large	0.1–0.2	Sillitoe *et al.* (1975)
Colavi-Canutillo	Bolivia	Stratabound, subvolcanic	16	0.45	COMIBOL Internal Report
Dachang	Guangxi, China	Limestone replacement	50–100	0.5–1.0	Under development by CNNC
East Kemptville	Nova Scotia, Canada	Polymetallic stock-work	56	0.2	Under development by Rio Algom: 5400 t/year Sn by 1985
Greenbushes	Western Australia	Pegmatite	28	0.11 + Nb, Ta 0.54 Sn eq.	Under study by Greenbushes
Hub	Czechoslovakia	Greisen	30–40	0.3 + 0.2% WO_3	Potential open-cut mine
Japo	Bolivia	Subvolcanic vein	25	0.35	Personal communication, COMIBOL
Lost River	Alaska, USA	Skarn-altered greisen	33	0.27	R G Taylor (1979) and ITC (1982). Possible open-cut mine
Monono	Shaba, Zaire	Decomposed pegmatite	100?	0.02	Sainsbury (1969). Potential open mine. Economics doubtful
Morococala	Bolivia	Subvolcanic vein system	17	Unspecified	COMIBOL (1982)
Potosi Hill	Brazil	Deeply weathered greisen	Undisclosed	Undisclosed	Under evaluation by Brascan/BP Presently worked in its weathered parts
Renison	Tasmania, Australia	Limestone replacement	17	1.07	Underground mining by trackless methods
San Florencio	Bolivia	Subvolcanic vein system	10–20	0.8–1.0	Rich (1980) Serious metallurgical difficulties
San Pablo	Bolivia	Subvolcanic vein system, "porphyry tin"?	100–1000	0.2–0.5	Sillitoe *et al.* (1975)
Santa Fe	Bolivia	Subvolcanic vein system	24	0.29	Personal communication, COMIBOL (1984)
Taronga	NSW, Australia	Quartz-filled fracture system	37	0.15 + minor Ag	Feasibility completed, uneconomic at present prices
Uis	Namibia	Altered pegmatites	33	0.14 + Ta	In production
Yinyan	Southern China	"porphyry tin"?	Unstated	0.3–0.4	Guan Xunfan *et al.* (1984)

TABLE 4 – Possible impact of the development of selected large primary tin deposits on the tin supply of market economy countries

Mine tonnages (Mt ore)	Mine life (years) by Taylor's rule	Ore throughput (t/year)	Tin recovered* (t/year)	Projected share of tin consumption†(%)
50	17.3	2 900 000	6 900	3.7
100	20.6	4 850 000	11 600	6.5
300	27.1	11 000 000	27 000	16.1

* Sn recovery based on: ore grade = 0.4% Sn; recovery rate = 60% Sn recovered = 0.24% Sn (0.4 × 60%).
† tin consumption of the market economy countries by the year 2000 projected at 186 000 t/year Sn.

will increasingly move away from alluvial operations and more into highly mechanised, modern, cost-effective hard-rock mines.

Considering the enormous and easily exploitable primary tin resources of China, it is probable that, before long, China will be once again a major tin exporter. Unquestionably the potential exists for China to produce in the order of 20,000 t/year of additional tin by the year 2000. From a global point of view, China will be able to supply 10% of the world's need for tin.

Countries with partially obsolete tin production facilities, particularly Bolivia and Malaysia, will be faced with the alternatives of either identifying new and low-cost tin resources or decreasing the share of tin supplied by them to the international market.

CONCLUDING COMMENT

Fundamental changes seem imminent in the world tin supply. A decreased tin demand and the vigorous entry of new producers, particularly Brazil and Canada, to the supply scene is jeopardising the effectiveness of the International Tin Agreement. A structural oversupply of tin will depress prices and unquestionably stimulate fierce competition among tin producers. High cost producers, particularly the small- and medium-sized ones, will be severely affected. A similarity with trends in the copper industry during the last few years can be drawn.

At the same time technology has become available for the exploitation of large,low grade, tin deposits. New and major hard-rock tin resources have been delineated in several countries in the last decade.

China, the United Kingdom and South Africa have seen an increased commitment to the exploration and development of hard-rock tin deposits. The growing perception in the industry is that large primary tin deposits will be highly competitive in the 1990s: a period when because of a weak or possibly nonexistent international tin agreement (ITA), there may be no price support for tin prices. Such a situation would permit the countries endowed with primary tin resources, and access to capital and technology, to develop a new generation of tin mines based on:

1. comprehensive exploration and evaluation of primary tin deposits (with may be a relatively costlyoperation);
2. economy of scale; and
3. modern, cost-effective bulk-mining methods.

THE TIN MARKET

At present tinplate manufacture accounts for about 35% of the total tin consumed. The future for this industry appears increasingly grim because modern tinplate manufacture needs only one gram of tin per square metre of steel compared with the two grams of tin per square metre necessary only a few years ago.

Tincans are now being welded instead of soldered and this, of course, eliminates the need for tin-based solder. Together with the introduction of aluminium cans and plastic and cardboard containers for frozen foods, this means a steady reduction in the demand for tin.

The often-stated assumption that the future of tin consumption lies in the industrial development of the Third World can be dangerously misleading. The lesson of development is that Third World countries tend to bypass.

It is tempting to speculate that if the Bolivian tin deposits were discovered today rather than a century agothey would probably be exploited from the start by bulk mining methods and be developed into very large, if lowgrade, tin projects rather than undergoing the ruthless high-grading to which they were subjected in the early stage of their exploitation.

Yet exploration for large hard-rock deposits is still in its infancy. Only a fraction of the capital spent, for instance, in the exploration for porphyry copper or massive volcanogenic base-metals has been spent on hard-rock tin exploration.

This situation is changing. In the last decade, systematic well-capitalised exploration for primary tin deposits has taken place in Australia, Canada, Brazil, Europe and Saudi Arabia. There has been less activity in Africa. In South East Asia (with the exception of China) a preoccupation with alluvial tin mining, a lack of risk capital and difficult tropical rainforest conditions have diverted attention away from exploration for this type of deposit. Bolivia, the western hemisphere's main hardrock tin producer, presents a unique case. Because of Bolivia's economic situation the exploration for new, and large, tin deposits is almost nonexistent. This is unfortunate because rent advances in the detection of blind or poorly outcropping orebodies appear eminently applicable to the well-exposed unforested areas of the Bolivian tin belt.

Considerable hard-rock tin resources mayexist in Indonesia, Thailand and, to a lesser extent, Malaysia. However, only 2% of the still enormous tin production of these countries comes from primary deposits. With alluvial operations, both offshore and onshore, becoming increasingly deeper and requiring very high capital costs, it is possible that the respective governments will reconsider and revise their kexploration and development priorities. An important factor in such revision will be the progressive and irreversible decrease in the production of small tin producers, such as gravel pumps in Malaysia and suction boats in Thailand. These producers face a severe period of low tin prices, stiffer government and environmental regulations, and depletion of easily won reserves. The progressive decline of the role of small and undercapitalised operators is shown by the fall in production in Zaire and Nigeria during the last decade. Their cumulative production dropped from 12691 t in 1972 to 3454 t in 1983, and continues to decrease. Other countries where operations are small and undercapitalised are expected to follow the same fate.

On the other hand, countries which are increasing their tin production in spite of depressed tin prices are the UK, South Africa and Peru, where modernly planned new mines have gone into production. Canada, with the commissioning of Rio Algom's new mine at East Kemptville, will emerge as a new tin producer, capable of meeting 3.7% of the demand for tin from market economy countries from this mine alone.

Tin production in Brazil, although almost entirely from alluvial sources seems all but unstoppable, and with production of 17000 t estimated for 1984 it represents some 10% of the production of the market economy countries. This is putting a formidable strain on the Sixth international Tin Agreement (ITA) and, now as in the future, the low cost Brazilian tin output will profoundly affect all tin producers.

It is significant that the tin-producing countries which are not members of the 6th ITA, with the notable exception of Brazil, are mainly hard-rock tin miners. Among non-members, China probably has the lowest mining costs.

Of the six countries within the ITA (see Table 2) only Australia is a significant hard-rock tin miner. Close to 90% of the 101 000 t of tin produced in 1983 by ITA members (and all of the 16000 t of the "tin of unspecified origin" which found its way onto the market) is of alluvial origin.

REFERENCES

Bettencourt,J.S.Damasceno,E.C.Fontanelli,W.S.Franco,J.R.M. and Pereira,N.M. 1981. Brazilian tin deposits and potential,ITC, Fifth World Conference on Tin, Kuata, Lumpur, 1981,(reprint, 69 pp).

Comibol,1982. Annual Report 80 pp (in Spanish)

ITC 1982. World Tin Mining; Operation, Exploration and Development. International Tin Council, London.

ITC 1984. Tin Statistics 1973-1983. International Tin Council, London.

Guan Xuntan et al.1984. A new type of tin deposit in China- the Tinyan porphyry tin deposit international Symposium on Geology of TinDeposits. Abstracts,pp 21-22, Chinese Academy of Geological Science, Beijing,China.

Mineral Yearbook 1982. US Department of Interior, Washington DC,p.234.

Mining Annual Review 1984. Mining Journal, London,p.142.

Plimer,I.R 1980. A review of the principal types of tin deposits in the world and models for their genesis in Tin in South Eastern Australia SMEDG Seminar, 1982,pp 1-3.

Premoli,C.1984. Exploration strategies for primary tin deposits. International Symposium on Geology of TinDeposits. Chinese Academy of Geological Sciences,Beijing,China (in Chinese).

Rich,P.J.H.1980. Future of tin as a tonnage commodity. Trans Inst Min Metall.Sect A.88,A8-A16.

Rongfu,P.Zhizong and Mingke,Fu 1983. A preliminary discussion on the sequence of solid mineral prospecting exploration and development. Bull Chinese Acad of Geol Sciences,5,13-16.

Sainsbury,C L 1969. Tin resources of the world.Bull US Geol Survey, 1301,55 pp.

Sillitoe,R.H.Halls,C and Grant,J.N.1975. Porphyry tin deposits in Bolivia. Economic Geol.70,913-27

Taylor,H.K. 1978. Mine evaluation and feasibilities studies. In Mineral Industry Costs (Eds J R Hosking and W R Green), Northwest Mining Association; Spokane,USA,pp 1-17.

Taylor,R.G. 1979. Geology of Tin Deposits, Amsterdam, Elsevier,543 pp.

RAW MATERIALS FOR STEEL PLANTS IN INDIA - CURRENT STATUS AND STRATEGIES FOR FUTURE DEVELOPMENT

G S Khuntia
Rourkela Steel Plant
Rourkela, Orissa
India

ABSTRACT

The principal raw materials used in steel making are iron ore, coal, limestone and dolomite. Raw materials constitute 40-45% of total input cost in steel making and therefore play a vital role. Highest productivity and economy in steel making is achieved by stringent quality of raw materials by improvement in processing and beneficiation techniques to prepare close size products, adoption of raw materials preparation process like ore bedding, blending, screening facilities both at Mines and Steel Plants. The paper highlights the adverse effect of silica and alumina in raw materials, various consumption rates, present and projected consumption pattern by 2000 A.D., quality stipulations, etc. The paper provides distribution of raw materials resources in India, linkage plan of ores and fluxes to steel plants, agglomeration and raw material preparation facilities, computer application in ore body modelling.

The paper also charts strategies for future development of bigger Mines, solution to problems of sizing, higher gangue content, increasing cost of raw material preparation, development of a specialist organisation for comprehensive programme for detailed exploration, exploitation, Mines construction and specialised management of raw materials production units.

INTRODUCTION

The principal raw materials used in steel making are iron ore, coal, limestone and dolomite, while the other raw materials used in small quantities are manganese, kyanite, quartzite, bauxite etc. Raw materials constitute 40-45% of the total input cost in steel making and, therefore, any progress made in controlling the cost of raw materials fed to steel works as well as its specific consumption per tonne of hot metal produced, shall substantially affect the economy in steel making. For improving the productivity in steel plants, the metallurgical process has demanded stringent quality of raw materials, improvement of processing and beneficiation techniques in the mines and adoption of new technologies like ore-bedding, blending and screening facilities both at the mine and the steel plants. Silica and alumina are the main gangue material in Indian iron ore, limestone, dolomite and metallurgical coal and these constitute slag as much as 500-600 kg per tonne of hot metal produced. For producing steel at the lowest production cost, the consumption rate per tonne of hot metal produced in respect of various raw materials should be aimed at their lowest, in addition to producing lowest slag volume.

CONSUMPTION OF RAW MATERIALS

For production of one tonne of hot metal in the blast furnace, the following quantities of raw materials are consumed on an average, though the figures vary from steel plant to steel plant in India and Table 1 indicates a comparison with Japanese steel mills.

TABLE 1

Raw Materials	Qty. per tonne of hot metal (Kg)	Japanese steel mill consumption rate(Kg)
Iron ore	1600-1700	1600
BF grade limestone	400	
SMS -do-	150	200
BF Dolomite	120	-)Serpentine
SMS Dolomite	60	-)consumed
Coke	860	400

The present consumption of raw materials as well as the projected consumption by the end of 7th Plan (89-90) and by 2000 A.D. presuming 24 million tonnes steel production by 2000 A.D. is furnished in Table 2 below (as per study by National Council of Economic Research with 4% annual growth).

TABLE 2

Raw Materials	Present consumption (MT)	Projected consumption (1989-1990) (MT)	Consumption by 2000 A.D. (MT)
Hot metal production	9.6	14.9	28.9
Iron ore	15.6	25.4	48.5
Limestone	4.7	8.0	16.0
Dolomite	2.0	3.5	6.8

PROBLEMS OF IRON ORE

Reserves

India is endowed with vast reserves of iron ore of good quality and the estimated reserve is furnished as under:

High grade direct shipping haemetite ore	11,470	million tonnes
Low grade magnetite ore	6,103	million tonnes.

Since the iron content in magnetic deposit is low, arouund 35-40% and it requires eleborate processing operation for enriching to a level of 60-65%, the cost of production of iron ore with 60-65% Fe is substantially higher compared to haemetite deposit, where the operations are much simpler and merely by size reduction and washing it is possible to produce iron ore lump and fines for supply to steel works. The above iron ore reserve is adequate for meeting the requirements of steel industry of India for the next 100 years at the rate of 75-80 MT per annuam, presuming 30 MT export. The five distribution of iron ore is in principal zones as under:

TABLE 3

Zone	State
A	Bihar, Orissa.
B	Bailadila, Rowghat and Dalli-Rajhara of M.P. and Surajgarh of Maharashtra.
C	Bellary & Hospet in Karnataka
D	Goa and Redi areas
E	Kudremukh, Bababudhan in Karnataka.

Quality

Apart from large reserves, India has very high quality iron ore averaging 62-63% Fe, though the ore from Bailadila is about 66% Fe, one of the best in the world. The quality of iron ore consumed in Indian steel plants is as under:

TABLE 4

Steel Plant	Fe%	SiO_2%	Al_2O_3%
DSP	60-62	2.5-3.5	5.5-6
IISCO	60-62	2.5	4 - 6
TISCO	63-65	$SiO_2+Al_2O_3$ =	6% max.
RSP	60-62	2.5-3	4-4.5
BSP	62-64	3-3.5	3.25-3.75
BSL	62-64	1.4-3	4-5.5

Linkages

Most of the steel plants have captive mines in iron ore, limestone and dolomite sector, while coking coal is purchased from Coal India Ltd. The raw materials linkage to various steel plants is furnished in Table 5 in the next page.

TABLE 5

Raw material linkage-source to steel plants

Steel plant	Iron ore	Limestone BF	Limestone SMS	Dolomite BF	Dolomite SMS
RSP	Barsua,Kalta	Purnapani	Satna	Birmitrapur	Baraduar
Bhilai	Dalli-Rajhara	Nandini	Nandini	Hirri	Hirri
Bokaro	Kiribure, Meghahatuburu	Bhavanathpur	Kutesvar	Tulsidamar	Bhutan
DSP	Bolani, Barajamda	Birmitrapur Purnapani	Satna	Birmitrapur	Bhutan
IISCO	Gua,Manoharpur	Birmitrapur	Satna	Birmitrapur	Baraduar
TISCO	Noamundi,Joda	Birmitrapur	Satna	Gomadih	Gomadih
Visakha-patnam	Bailadila	Jaggayyapet	Badhan-pur	Machkot Kotmir	Karepalli

The above mentioned mines are linked to steel plants based on available information in the past on the prospecting of various iron ore and limestone/dolomite deposits. With the increased exploration efforts and information available now, a re-thinking will be necessary to rationalise the present linkages of the captive mines to the steel plants taking into consideration the proximity of such deposits from the steel plants and various other techno-economic considerations.

Steel production

The projected hot metal capacity and expected production by the end of 7th Plan and estimated capacity by 2000 A.D. have been indicated in Table 2.

The requirement of about 50 MT of iron ore by 2000 A.D. can be met from the large reserves of iron ore in the country. Even with the increased level of export to 30-35 MT it is expected that these reserves would last for about 70 years or more.

Iron ore feed - product mix

Iron ore feed in the blast furnace constitutes lump, and ore fines which are agglomerated in the form of sinter or pellets. The earlier steel plants were designed for iron ore lump feed, while the recent plants like BSL/BSP/RSP/DSP/Vizag are designed for utilization of agglomerated fines in the shape of sinter with improvement in steel technology. It has been broadly estimated that for every 1% increased use of sinter, it raises the output by 3.5% and coke consumption is reduced by 3%. This has resulted in progressive increase in the use of sinter in the blast furnace burden today to achieve high productivity and cost efficiency. In Japanese steel mills, which are known for their high efficiency and low cost of operation, the average feed constitutes about 80-90% of agglomerated fine, which comes to about 70-80% sinter and direct lump ore consumption ranges from 20 to 25% only. The proportion of lump and sinter used as well as sintering facilities in Indian steel plants is as under:

TABLE 6

Plant	Lump %	Sinter %	Annual installed capacity (MT)
IISCO	100	0	-
Durgapur	70	30	1.52
TISCO	60	50	0.80
Bhilai	60	40	3.37
Rourkela	60	40	1.20
Bokaro	30	70	4.1
Visakhapatnam	20	80	

Although there are three pellet plants i.e. Noamundi mine of TISCO, Chouwghule & Co at Pale, Goa, Mondavi, Goa and one pellet plant of 3 MT annual capacity at an estimated cost of Rs. 107 crores is approaching completion at Mangalore for utilising Kudremukh iron ore fines for agglomeration, pellet making in India is not finding popularity due to high costs and non-availability of inputs like furnace oil and electricity.

Ore preparation

In present day blast furnace practice, ore preparation plays a vital role in productivity of blast furnaces requiring intensive beneficiation and materials handling process for achieving the following advantages:

- increasing iron content
- removal of gangue/undersize material
- meeting cloe size requirement
- homogeniety in chemical composition
- agglomeration of fines for improved economics in iron ore mine workings.

Quality

For 1% increase in Fe, there is 1.25 - 2.5% decrease in coke consumption, 25-30 kg reduction in flux consumption per tonne of hot metal. In Japanese practice, the Al_2O_3 in the iron ore feed is restricted at 1.8% (max) and critical values are stipulated for S,P, alkalies etc. In Soviet practice, by suitable raw material preparation it is possible to restrict feed with Fe variation ±0,15%, SiO_2 & Al_2O_3 variation ±0.33% resulting in saving coke consumption rate by about 4%. Iron ore consumption per tonne of hot metal averages to 1700 kg, both in form of lump and fines.

Size

The size of the lump and fines in integrated steel plants in India at present is 10-40/50 mm and 100 mesh -10mm, while in the Japanese plants it is 8-30mm (3% reduction in coke consumption is achieved). The present trend world over is to have a ratio of top size: bottom size lump ore = 3:1 (preferably 8-25mm). Bhilai, Bokaro and Rourkela Steel Plants have decided to reduce the iron ore lump size to 8-30mm.

Sinter feed

For achieving maximum productivity and minimum cost in steel products, in Japanese practice the sinter charge in blast furnace varies from 70-100% with -100 mesh iron ore fine 40% tolerance, whereas in Indian plants such tolerance is less than 10%. This has resulted in restriction in high grade iron ore fines "blue dust" known to be very rich in Fe% but with -200 mesh size. Experiments in TISCO plant could establish the use of blue dust upto 30% in sinter and similar results have been achieved in experiments conducted by Research & Development Centre, Ranchi for utilising -200 mesh microfines to about 20% with or without addition of lime in sinter making in case of Barsua Iron Mine of Rourkela Steel Plant.

Undersize tolerance

For achieving reaction and gas permeability 5 mm tolerance in lump ore is targetted at 1% and at 3% in sinter-feed in Japanese steel mills, while such tolerances are much higher in Indian steel plants.

Reduction/degradation index

The index in Japanese steel mill is as under:

RDI	Lump	Sinter
at-5mm	90% max.	40% max.

Similarly decrepitating ore is not used in blast furnace and in case of lump ore, decrepitation index (-5mm at 3%) is insisted on in Japanese plants.

Ore preparation with washing & wet screening

For achieving the desired proportion of lump and fine and improvement in quality and for supplying close range calibrated size products, most of the mechanised iron ore mines attached to steel plants have introduced multi-stage crushing -screening and a few with washing facilities. The grade of the ore washed in some of the plants is furnished below:

TABLE 7

Mine	Product	Fe	Quality SiO_2	Al_2O_3	Size		
Bailadila	Lumps	67	2.5	3	-150mm	+	9.5mm
NMDC Ltd	Fines	65	-	-	+65mesh	-	9.5mm
(For export)							
Kiriburu	Lumps	63±1	1.5±0.4	4.0	+10mm	-	40mm
BSL	Fines	62.5±0.3	3.5	5.0	+100mesh	-	10mm

In order to cater to the need of lumps at 8-30mm, it will be necessary to introduce tertiary crushing-screening facilities with wet process in most of the captive mines attached to steel plants very shortly.

Blending of ore

Three or four benches are operated at a time for feeding ROM ore in the quarry to the crushing-screening plant for achieving the above objective. In order to supply uniform quality ore most of the mines have provided or planning to provide the following facilities:

- rigid quality control in mine operation by operating 3-4 benches at a time.

- adequate provision of blending and stockpile with facilities for stacking and reclaiming before despatch at mine heat

- adequate provision of stockpile with modern blending arrangement in the steel plant in the form of stacker and reclaimer.

The desired level of quality control is achieved through:

- adequate core drilling to realise the desired level of accuracy in ore reserve estimation and computerised ore body modelling with mine production planning

- implementation of such production scheduling through effective dumper despatch programme from individual mine bench and adequate sampling and monitoring arrangement

- adequate stockpiling facilities by provision of stacker/reclaimer for providing ore with uniform chemical and physical characteristics, provision of ore bedding and blending facilities in the steel plants with stacker/reclaimer and provision of block-rake system of 4500/7500 tonnes utilisation of micro-fine/blue dust.

The blue dust constitutes 6-10% of the total reserves of Iron ore and is very rich in Fe content, but conforms mostly to 100-200 mesh size. By washing process most of the blue dust will be lost to tailings disposal system and therefore dry circuits have been provided in some of the mechanised iron ore mines like Barsua/Dalli/Kiriburu-Meghahatuburu. For achieving maximum economics in the ore production and for conservation of mineral resources it is desirable that such products are utilised in sinter making by addition of lime in the sintering process.

LIMESTONE

The consumption of limestone both of BF and SMS grade as of present and future is as under:

1984-85	5.00 MT
1989-90	8.00 MT
2000 AD	16.00 MT

The reserves are 6300 MT for blast furnace grade and 1000 MT for SMS grade.

Table 8 below indicates the captive mines, existing capacity, the reserve of ore and fluxes and quality of ore and fluxes supplied.

TABLE 8

Steel Plant	Captive mine	Existing capacity (finished product) (MT)	Reserve (MT)	Quality Chemical	Physical
Bokaro Steel Plant	Kiriburu Iron ore Mine	4.2	340	Lumps 63% Fe SiO_2 61.6% max.	10-40mm
	Meghahatuburu Iron ore mine	4.2		Fines Fe 62% SiO_2 3% Al_2O_3 5-5.5% max.	+100mesh -10
	Bhavnathpur Limestone Mine (BF)	1.45	100	CaO 42%	25-80mm
	Kuteswar lime-stone mine (SMS)	0.40	93	CaO 50% TI 5% mx.	30-60mm
	Tulsidamar Dol. Mine	0.60	20	CaO 31-32% MgO 21-16% SiO_2 01.2-4% Al_2O_3 3-2.5%	25-80mm
Rourkela Steel Plant	Barsua Iron Mine	2.1	470	Lumps Fe-61% Al_2O_3-4% SiO_2-2.5%	10-50mm
	Kalta Iron Mine			Fines Fe-60% Al_2O_3-4.5% SiO_2-3%	+65mesh -10mm
				Fe-63%	10-50mm
	PLDQ(BF lime-stone)	0.62	30	TI 11±1% CaO 44%	25-50mm +60mm 10%max.
	ILQ, Satna (SMS Limestone)	0.15	50	TI 5.5% CaO 50%	40-80mm
	IDQ,Baraduar (SMS Dolo.)	0.24	65	TI 5% max. MgO-20%	40-80mm

Bhilai Steel Plant	Dalli-Rajhara group of mines (Iron ore) including manual mine	5.4	325	Lumps Fe-63% SiO_2 2.99% Al_2O_3-3.27%	10-50mm
				Fines Fe-62% SiO_2-3.51% Al_2O_3-3.27%	-10mm
	Nandini Limestone mine. (a) BF L.st.	1.4	190	CaO-41.75% MgO-8% SiO_2-5.1% Al_2O_3-3.18% AI-8.28%	+6mm/ +12mm upto 50mm
	(b) OH/RMP grade Limestone			CaO-49.52% MgO-2.56% AI 4.95%	+50mm -100mm
	Harri Colomite Mine	0.60	42	Lumps AI 3.524% CaO-50% MgO-20% SiO_2-3%	50-100mm
				Chips AI 6-7% CaO-29% MgO-19% SiO_2-5-5.5%	-50mm
Durgapur Steel Plant	Bolani Iron Ore Mine	2.6	460	Lumps Fe-60-63% Al_2O_3-5.5% max.	10-50mm
				Fines(Unwashed) Fe-56% Al_2O_3-6% SiO_2-3.5%	-10mm
Indian Iron & Steel Co Ltd	Gua Iron Ore mine	2.4	230	Fe-61% SiO_2-2% Al_2O_3-4%	12-75mm
	Chiria Iron	0.50	1869		
	Chasnala Jitpur Ramnagar Colly.	1.0	70		
Tata Iron & Steel Co Ltd	Noamundi Iron Ore Mine	2.5	-	OH Gr.Fe Lumps	+50mm
				Fe-66% Al_2O_3-2.2%	10-50mm
				Fines Fe-63.5% Al_2O_3 & SiO_2-6% max.	+100mesh -10mm.

Joda Iron Ored mine	1.0	-	Lumps Fe-66% Al_2O_3-2.2%	10-50mm
Gomardih	0.60	-	TI-8% max. MgO-19% min.	25-75mm
Joda West Mn. Mine	0.05	Mn. ore.		
Hathibari Limestone Mine(BF)	0.15	-	TI-8-12% M g O - 6 %	-75mm
Chromite Mine at Sukhinda	-	-	-	-

Out of 1000 MT of SMS grade limestone, about 600 MT is available in Kuteswar area upto 100 metre depth.

Linkages

Most of the steel plants except TISCO/IISCO/DSP, who receive their supplies from Birmitrapur, have their own captive mines and the linkage has already been furnished in Table 8. Similarly, BSL/RSP/BSP have their own captive mines for supply of SMS grade limestone but they are self-sufficient to about 60-70%. Expansion programme, through introduction of mechanised mining and crushing-screening facilities, is being planned for Kuteswar (1.5 MT finished product per annum) and ILQ, Satna (7 lakh tonnes per annum) and Nandini manual mine/Shahpura Limestone mine (2.5 lakh tonnes per annum).

Since good quality of SMS grade limestone cannot be achieved by mechanisation in quarry as it will result in dilution of the feed to the crushing-screening plant, it is planned to introduce 50-300mm boulder to be produced in the quarry manually and feed the same by mechanised transport to crushing-screening plant to achieve a close range product with under-size/oversize limited to 10%. However, in the area of limestone, 'both BF and SMS grades, though the present size is 25-75mm except BSL (30-60mm), most of the steel plants are planning to introduce 10-50mm size limestone. The balance quantity of SMS grade limestone is purchased from sources at Katni/Jukehi/Satna/Rewa area from M.P.

Constraints

Inadequate SMS grade reserve

Except Kuteswar deposit the reserve of SMS grade limestone in the country is quite limited. Systematic exploration is required in undermentioned areas where SMS limestone with SiO_2 less than 1% has been reported.

a) Himalayan occurrences in Dehra Dun and Tehri districts of Uttar Pradesh, Himachal Pradesh and Meghalaya.
b) West Rajasthan.

Quality - High Silica in SMS Limestone

SMS limestone supply from M.P. is characterised by high silica of about 3% compared to standard practice of 1% silica in SMS limestone used in Japanese mills. It is, therefore, necessary to identify and develop sizeable supply of this SMS grade limestone.

Small manual quarries

As the SMS grade limestone requirement is not very large, manually operated quarries are developed as captive sources as well as in the non-captive sources. Due to engagement of labour on contractual basis in non-captive sources, the supplies are very often disrupted due to industrial relations problems.

Wastage of fines

The undersize i.e. -25mm fraction that will be generated in the process of crushing-screening, will constitute as much as 30% and therefore will pose a problem in the economics of mine working unless it finds a suitable market in neighbouring cement industries.

Similarly BF grade limestone is often associated with SMS grade limestone and therefore despatch of this BF grade limestone should also improve for economic working of such SMS deposit.

Recommended steps

a) Development of one or two large capacity (1.5 - 2 MT product per annum) mines in M.P. by mechanised means for ensuring supply in quality and quantity.

b) Identification of superior grade i.e. low silica deposits in Himalayan Belt for supplying to different steel plants for blending with the existing supply of SMS grade limestone.

c) Studies to be undertaken for improvement in quality of SMS grade limestone from M.P. through proper processing.

DOLOMITE

Consumption

The consumption of Dolomite is about two million tonnes per annum in the Steel Plants both in the process of steel making as well as for refractory purposes. The norm of consumption of Dolomite varies from plant to plant with 5--200 kg of BF grade and 40-60 kg of SMS grade Dolomite per tonne of hot metal produced.

Reserves

Reserves of BF grade and SMS grade Dolomite is about 1426 MT (updated to 1976) distributed in three zones in the country shown in the next page.

Zone	Total reserves (MT)
A. Bihar & Orissa	581.25
B. Central India & Andhra Pradesh	525.01
C. Karnataka, Tamil Nadu & Andhra Pradesh	202.87
Total	1309.13

SMS grade Dolomite is produced and supplied from Captive mines of Steel Plants at Baraduar (RSP), Hirri (BSP), and Tulsidamar in a limited quantity (BSL), while BF grade Domomite is supplied to the steel works from non-captive sources at Birmitrapur and other captive sources like Tulsidamar (BSL) and Gomadih (TISCO).

Quality

The desired specification of Dolomite for consuming industries in India is as under:

Industry/Gr.	CaO+MgO% (Min)	MgO% (Min)	SiO_2% (Max)	A.I.%
Blast Furnace/ Sinter/Pellets	48	18	4	8
Steel making shop	50	40	2	4
Refractory making	52	22	1	2

A number of areas of refractories have been identified by a technological planning committee and RDCIS, Ranchi, SAIL, have taken up a few works in the areas of refractories for development of tar bonded Dolomite bricks and high alumina refractories which will increase the life of convertor lining considerably and reduce the shut down period. SMS grade Dolomite except for BSP, for other Steel Plants meet their shortfall for the production from their captive mines, from Baraduar area where in addition to RSP mine, another 2 - 3 private mines are in operation, as well as from Bilaspur area.

Utilisation planning

While Baraduar, Hirri and Bilaspur area are meeting the major requirement of SMS and refractories grade Dolomite, the reserve being limited, additional exploration is necessary, in Bhutan/Jayanti and Dehra-Dun area, known to have huge reserve of low silica superior grade Dolomite, mining operations in Dehra-Dun both low silica limestone and Dolomite is being carried out in limited scale due to environmental problems. But some steel plants (BSL, RSP, TISCO) are getting low silica Dolomite from Bhutan/Jayanti. The grade of low silica Dolomite lump are compared to that supplied from Hirri/Baraduar is as under:

	CaO%	MgO%	SiO_2%	R_2O_3%	A.I.%
Hirri/Baraduar	29-30	20-21	2-2.5	1.5	3 - 3.5
Jayanti/Bhutan	30	21	1.0	1.0	2

Detailed exploration is necessary in the deposits at Jayanti/Bhutan. Although there are indication that these deposits are quite extensive and have vast potential, SAIL with WBMDC are jointly developing infrastructure facilities for developing mines at Jayanti. Visakhapatanam and Vijaynagar Steel Plant is planning to obtain Dolomite from Khamam and Bagalkot area respectively.

Constraints

Small manual quarries

As the requirement of Dolomite is comparatively small, captive mines of lower capacity are attached to Steel Plants and the shortfall in supply is met from non-captive sources having very small capacity mines with contract labour. With the prohibition of contract labour engagement in mining operation, this is causing serious problems in the operation of such small manual quarries in the private sector and even in public sector. Therefore, a comprehensive programme for development of Dolomite mines with adequate high capacity of production with the concept of boulder mining (50-300mm size) and mechanised crushing and screening facilities will go a long way in sorting out the above problems for regular and consistent supply, both in quantity and quality in steel works.

Undersize

A lot of undersize (-40/-25mm) to the tune of 20-25% is currently not having a market, thrown as waste and affecting economy in Dolomite production. Therefore, alternative utilisation of undersized Dolomite in sintering making and in other industries may improve Dolomite mining economy.

Recommended steps

Through opening a large capacity mechanised mines to solve the above constraints, it may be better to have 2 - 3 mines having higher capacity of more than one million tonnes product per annum and opened at Hirri/Baraduar/Tulsidamar area with boulder mining for feeding Dolomite boulder of specified quality for crushing and screening. By this, it will be possible to supply good quality as well as quantity of both BF grade and SMS grade Dolomite from Tulsidamar/Hirri/Baraduar to steel works.

STRATEGIES FOR FUTURE DEVELOPMENT

Problems in quality of flux

Sizing

Sizing of flux material is a major problem and steel plants even with sintering facilities are hesitant to purchase fines from the mines because of fear of contamination and high percentage of insolubles. It is necessary to carry out systematic study to overcome this problem so that both lump and fines can be produced in respect of limestone as well as dolomite, both of BF and SMS grade so that flux material can be produced and supplied to steel works economically. Solution to this lies in boulder mining as well as crushing and screening facilities and utilisation of lump as well as fines.

High Acid Insolubles

High acid insolubles in fluxes is causing serious problems. Blending in a limited way in steel works is a short-term solution while long-term solution calls for detailed exploration and development of low silica limestone as well as dolomite mining in Hirri, Mossorie, UP for Limestone and Jayanti/Bhutan, Khamam for dolomite so that such fluxes of superior grade received from distant places can be blended with existing supplies from the captive mines for effective blending and enrichment.

Costs

Preparation of ore and fluxes to meet the stringent physical and chemical parameters at the steel works, requires the installation of sophisticated facilities at the mines, employment of highly skilled personnel and other inputs. Although quite a lot of work has already been done in this area, but it is very important that much more inputs is required to be fed in the captive mines of the steel plant for better preparation of the raw materials with reasonable increase in additional cost for achieving greater benefits that flow out of the use of better prepared raw materials. In respect of raw materials that will be fed from non-captive sources, a system of consultation, inspection, element of prices related to specific improvement etc. increase in the cost within a reasonable limit can be introduced.

Organisation

Modern blast furnace technology practices demand better prepared ore and fluxes so that efficiency and productivity of blast furnace is determined more on what is done on raw materials outside the blast furnace than what is done in the raw materials inside the blast furnace. This calls for highly competent management to ensure consistency in chemical and physical specification of the ore and flux as well as good measure of flexibility to meet the changing needs and to give opportunity for blending of ore and fluxes received from other sources when necessary.

SAIL has already created an organisation CRMM, Ranchi, for controlling, guiding operations for carrying out innovation, allocation of resources for operating screening and blending facilities, exploration facilities. Since the target of steel production by the turn of the century is set at about 24 MT, it calls for formulation of integrated and comprehensive programme for exploration and exploitation of the deposits, construction of mines and their operation. The development of a specialised organisation for the entire raw material resources under SAIL may resut in feeding rich and consistent quality and quantity of raw materials to the steel works.

Environment & Ecology

Regular monitoring of following areas is necessary for maintaining clean environment:

a) Water pollution

b) Air pollution

c) Noise pollution

d) Massive tree plantation for arresting soil erosion and deforestation

e) Reclamation of mined out area.

ACKNOWLEDGEMENT

The author is thankful to the Managing Director, Rourkela Steel Plant, for his kind permission to present the paper. The opinions expressed in the paper are that of the author only.

REFERENCES

1. Gupta, P.C. and Rao, C.S.V. - Raw Materials for Steel Plants: Round Table on Steel Industry for the next Decade, Feb 18-19, 1985.

2. Khuntia, G.S. - Mining of Raw Materials for Iron & Steel Industry of India, the present and future, on Indian Mining - 1984, Journal Mines Metals & Fuels, Spl. Number, 1985.

COGNATE ISSUES

MINING IN DEVELOPING COUNTRIES - A VIEW POINT

Rajeshwar Thakore
Aryan Mining & Trading Corporation Ltd.,
Calcutta, India.

ABSTRACT

Mineral deposits are finite. Therefore, scientific and optimum utilisation of available mineral resources plays an important role in economic development. According to Chanakyas 'Arthashastra' "Mines are the source of treasury, from treasury comes the power of Government and the earth whose armament is treasury is by means of treasury and army". This shows that even in pristine days also, great emphasis was given to mining activities. With the growth of civilization, this should be of greater significance today.

With almost all countries stepping up their mining activity which has now become a base of the economic development of a country, it becomes imperative to carry on mining in developing countries in a most conscious manner with due regar d to cost effectiveness conservation, beneficiation, infra-structural facilities and safeguarding environment and ecology.

INTRODUCTION

After agriculture, mineral resources constitute the inherent strength of a country. Since the Industrial Revolution, the use of minerals has been equated with material success and progress of mankind and sine qua non of national prosperity. Prof. J.Cameron rightly said, "Minerals are stepping stones for the ascent of man; and mining and metallurgy are the architecture for that ascent and indispensable to the advancement of civilisation - the very spearhead of man's development."

A nation endowed with nature's bounty occupies a pride of place amongst the comity of nations and therefore optimal utilisation of available resources is as much strategically important as formulating an industrial policy of the country. It is heartening therefore, that the Govt of India too is in the process of formulating a National Mineral policy. More so, for developing countries who are always handicapped due to weak economic base and political polarisation as compared to the developed countries.

It therefore becomes imperative in the present stage of industrial development that the developing countries evolve upon a uniform and common policy which could forestall any exploitation of the less developed countries by their economically and industrially advanced counterparts.

The task, ofcourse, is not so simple. In fact, the developing countries are confronted with virtually a Hobson's choice in the matter of exploitation of their mineral resources. On the one hand, a price consideration is to keep the balance of trade healthy by earning foreign exchange, on the other, the countries' non-renewable resources are exported often unprocessed. Many advocate that the developing countries should not fritter away their natural resources but should export value added, processed, semi-processed and finished goods.

It is however an acknowledged fact that despite serious reservations, the developing countries without exception cannot do away without exports; the internal requirements being neither very buoyant presently nor visualised to be so capacious in future so as to really cause anxiety in the matter of exports. The responsibility of evolving a common strategy for exploitation of mineral resources in the developing countries becomes therefore all the more significant. Maximising exports at the cost of bailing out the best quality of minerals is a reflection of the helpnessness and the weakness of the developing countries should be regulated in a planned manner. If some pre-emptive measures are not taken now, a cut-throat competition may become more pronounced among the developing nations, providing long and effective handle to the developed importing countries to coerce them as much as possible, not only in respect of quality but in price also. The inscrutable laws of demand and supply should be understood thoroughly by the developing countries to combat the strategy of developed importing countries before the best of their mineral resources are wantonly squandered. What is therefore, urgently required is not only to bank on exports alone but to simultaneously develop and broadbase country's own industrial development assimilating the best available technological knowhow and ensuring utilisation of not the best quality of mineral alone but of average grade too. Even the export quality of minerals should not be the best available grade in the country but of mixed and average grade. This is only feasible when exporting countries formulate a uniform strategy from a common platform and forum.

However, in practice such a utopian dream may never materialise. The concept and ideal of consortium hardly ever fructifies as has been seen in the case of Iron ore, Jute etc It naturally follows that economic exigencies and political priorities lead to situations, where long term common interests and advantages are glossed over in favour of immediate gains. It is important for the developing nations to chart their own exclusive strategies to carry on their mining activities, keeping in view the immediate needs of the present and the interests of the posterity. From this point of view, certain broad strategies could be outlined.

COST EFFECTIVENESS

The future development of mining in developing countries is crucially dependent on the cost of production. In an era of great competition between oil and coal and when many other minerals are being replaced by other substitutes because of changing technologies, economics has to be identified as a problem inviting the highest priority. Prof.Atkinson put it very succinctly "Mineral prices are usually the most critical input in determining profitability, they are market demand determined rather than cost-determined and therefore extremely unpredictable. The cyclical nature of mineral prices is well known. They are stable only when supply and demand are in balance and expected to remain so. Any imbalance can lead to a disproportionate price reduction". In this context, it can also be said as far as Manganese ore and Iron ore production in India is concerned that there is a massive surplus capacity with many potential projects already in the pipe line. It is, however, to be remembered that excessive generation of surplus may bring down the prices over a period of next 10 years, obviously due production well in advance of actual demand. It is therefore essential to look at the demand forecast for exports as well as internal consumptions that production is not augmented to a level that instead of accruing benefits to the country, it falls in the trap of diminishing returns. The developing country cannot afford to block huge resources on infructuous investment which could have been gainfully utilised in the sectors needing them most.

If this is lost sight of, the developing countries in particular will face crises from more than one angle. Excess production over actual demand shall haul down the prices to an unremunerative level, unwarranted stock-piling shall result in drying up the liquidity. The resultant resource crunch can dim the long term perspective and planning. The President of a successful mining Corporation has remarked in the context of difficult world economic situation that "Survival during this period (the next 20 years) will be possible only for those who are truly the fittest companies in every industry and in e very country in the world". Here, the fittest "company" shall symbolize the entire infrastructure from winning of ore to the end of the line to consuming centres. How much cheaper, better and assured quantum of raw material is made available to the manufacturing units shall determine the survival of that producer company and on the global level that particular country.

In this context, one can cite the example of North American Iron ore Industry. According to Robert F. Anderson, Chairman of M.A.Hanna Co., Iron ore producers must reduce their cost substantially to compete in the global iron ore market and goes on to say that "if we fail to sufficiently reduce our costs, we can surely expect domestic steel producers to reduce their reliance on North American iron ore by importing ore from overseas or shutting down the hot end of the processes and importing foreign slabs for processing".

It is absolutely essential therefore that particularly developing countries do excercise effective control to keep the production costs under check during periods of inflation by taking to new mining techniques, determining the optimum size of mines, relationship of size to mining efficiency, improvement in productivity, economic results by judicious mechanisation and automation of production and exploitation of mineral resources at a rate depending on the existing local conditions.

INFRASTRUCTURE DEVELOPMENT

An adequate network of communication and transport is a pre-requisite for optimum utilisation of mineral resources. Mineral deposits are generally located in interior and far-flung areas well inside hinterland which call for proper outlet by economic and cheapest modes of transportation. Long haulage over land, either by road or rail is becoming a very costly proposition because of the fact that each stage of transportation is managed by individual agencies, ports being the last link as far as exports are concerned. Every organisation in the transportation link, obviously, is managed by separate corporate entities desirous of earning some profit out of their activities. Hence, multiplicity of profit margin adds upto increase the value of the material. Therefore, instead of different agencies, if the entire transportation work is carried on by one single entity, the incidence of profit element is confined to just one place thereby reducing the value of the material or in other words reduction in the total cost of the material. Increasing railway tariff, estalating port charges render the product uncompetitive in the international market. The table below gives an idea of the escalation in India:-

Railway freight rates per tonne on iron ore
(In Rupees)

	Banspani to Paradeep	Tornagallu to Madras
1979-80	60.90	51.50
1980-81	69.00	59.02
1981-82	84.70	71.60
1982-83	96.30	80.30
1983-84	118.00	98.20
1984-85	118.00	98.20
1985-86	129.83	108.00
%age increase over 1979-80	113	109

Here, it may be interesting to compare the railway tariff in Brazil. The iron ore mined at Carajas deposit of CVRD in Brazil has to move over a distance of 890 Kms in rake loads of 15-16,000 tonnes. The turnround time is reported to be 18 hours. The present freight is about US $ 4 per tonne which is likely to come down to US $ 2.50 per tonne when the mine is fully developed to produce 35 million tonnes of ore per annum.

Along with the mining activities, equivalent transport facilities, commensurate port capacities are to be built up so that large ships of current international standard could be berthed and loaded at a capacity available at ports of other exporting countries. The developing countries must plan to achieve international standards for berthing large vessels toretain their share in the global market. Unfortunately for India, there is a long way to go to achieve this capacity which is an extremely important factor to ensure faster growth of exports at a competitive international level.

Hence, verycareful planning is required to avoid slippage at any of the infrastructural phases in order to instil confidence among the importing countries, otherwise, it has been experienced several times that despite availability of graded, quality materials, due to inadequate infrastructural facilities, the deals are lost to countries with better equipment and allround facilities. The developing countries should engage in planned development which should take into the perspective future possibilities for the next 20 years.

ENVIRONMENTAL SAFEGUARDS

Mining comprises only a small fraction of the environmental damages as compared to agricultural and the demands of burgeoning populations which is evident from the following statistics of Indian conditions: "Out of every lost hectare of forest cover: 0.71% is by agriculture projects; 0.12% is claimed by river valley projects; 0.04% is by new industries, 0.02% by roads and communications and 0.11% by other miscellaneous purposes". It therefore becomes evident that mining activities do not play as destructive a role in destroying the forests and disturbing the ecology and environment as has been widely projected. Nonetheless, it is necessary that detailed planning State-wise, district-wise is undertaken to evaluate the mineral deposits quantitatively and qualitatively vis-a-vis ecology and environment so that vital mineral deposits are not left unexploited. Another way to reduce deforestationcan be to cut the trees on the existing leasehold first than from other periphery virginareas to meet the demands of railways, shipping, paper mills, furnitures etc. Mostly State Govt. Forest Corporations are undertaking deforestation in a systematic and planned manner. As such there can hardly be any difficulty in selecting priority areas where minerals occur. This can pave the way for striking a judicious balance between exploitation of strategic mineral deposits which are as much important to mankind as the upkeep of ecology and environment for the present as well as for the posterity.

CONCLUDING REMARKS

Developing countries should therefore plan the exploitation of mineral resources keeping in view the requirements of the present by careful cost-effective measures, by providing adequate infrastructure, by resorting to conservation of minerals through adoption of latest beneficiation technologies and last but not the least, not only by keeping the fragile ecology and environment intact but making it better for generations to come.

PETROLEUM STRATEGIES FOR THE THIRD WORLD COUNTRIES

V Raghavendra Rao
University of Nigeria
Nsukka, Nigeria.

ABSTRACT

Based on the criterion of petroleum resources, four groups of third world countries could be recognised, viz. those with surplus petroleum; those self-sufficient in petroleum; those partially self-sufficient in petroleum and those with no petroleum; these groups can further be sub-divided, on the basis of population, into those with big populations and those with small populations.

The more important strategies to be followed by these countries, for a healthy economic and industrial development include (i) providing adequate funds and favourable conditions for petroleum exploration; (ii) applying enhanced recovery methods in existing oil fields; (iii) better utilization of natural gas; (iv) development of other energy resources like coal, solar energy, hydroelectricity, nuclear energy and biogas; (v) establishing petroleum refineries and petro/coal chemical complexes; (vi) funding research and development in energy related projects and data storage and dissemination; (vii) encouraging conservation of crude oil and petroleum products; (viii) organising training programmes for meeting manpower requirements in the energy sectors; (ix) stock-piling crude oil and refined petroleum products for meeting crisis situations and (x) establishing units and "think tanks" for short and long range energy resources development planning and advising governments. Some of these strategies are common to all the groups of third world countries; each country has to adopt and choose the strategies relevant to its needs.

INTRODUCTION

The petroleum resources and the population are two of the more important parameters, that have a great influence on the industrial development and economic stability of any nation. Other factors that influence the inflow of foreign capital into third world countries are the attractive terms and conditions offered to foreign companies, and stable political and economic environments. It is very essential that all countries, third world countries in particular, should chalk out short and long range strategies for the development of their energy resources. This paper emphasises the strategies for the development of the petroleum resources.

CATEGORIES OF THIRD WORLD COUNTRIES

Four categories of third world countries could be recognised on the basis of petroleum resources. Group I: Those with surplus petroleum resources, for export; Group II: Those self-sufficient in petroleum; Group III: Those partially self-sufficient in petroleum and Group IV: Those with no petroleum resources. These could further be sub-divided, on the criterion of population, into (i) those with big populations and (ii) those with small populations. Examples for these groups and sub-groups are: for I,i): Nigeria; for I,ii): Kuwait; for II,i): Argentina; for II,ii): Tunesia; for III,i): India; for III,ii): Australia; for IV,i): Sri Lanka and for IV,ii): Mali.

PETROLEUM STRATEGIES FOR THIRD WORLD COUNTRIES

The following are some of more important petroleum related and other energy resources strategies that need to be adopted for a sound economic and industrial development of third world countries:

a) Intensifying petroleum exploration for the discovery of new reserves in sedimentary basins. Adequate funding, stable political and economic environments and attractive terms and conditions for foreign investment in petroleum exploration are needed for this purpose;

b) Applying enhanced recovery methods in existing oil fields. These methods can improve the recovery factor considerably, thus adding to the recoverable reserves;

c) Better utilization of existing natural gas resources. This will help in releasing crude oil supplies for critical industries like the petrochemical complexes and for export;

d) Development of other energy resources like coal, solar energy hydro-electricity, nuclear energy, biogas etc. This will also result in releasing more crude oil for critical industries and export.

e) Establishing petroleum refineries, petrochemical and coal based industrial complexes. These will help in the industrial development, creating new employment opportunities and stabilising the economies of third world countries.

f) Providing adequate funds for Research and Development in petroleum and other energy sectors and for establishing data storage and dissemination systems.

g) Encouraging practices to conserve crude oil and petroleum products.

h) Organising training programmes for meeting the man power requirements at all levels, in different energy sectors.

i) Stockpiling crude oil and petroleum products for meeting crisis situations like the war times and blocades of transport routes etc; and

j) Establishing planning units and "think tanks" for short and long range planning and for advising governments on energy related matters.

CONCLUSIONS

Each third world country has to adopt the strategies relevant to its petroleum resources position, its population and its special needs. Majority of the strategies mentioned above are relevant to all the groups of third world countries.

For those third world countries with no petroleum resources, special help from international organisations like the United Nations agencies, the World Bank and the International Monetary Fund etc. will be needed for finding new oil and gas reserves and developing other potential energy resources. The more developed third world countries like Brazil, Saudi Arabia, India etc should help other third world countries, in petroleum exploration and development.

Table 1: PETROLEUM RESERVES AND CRUDE PRODUCTION OF SOME THIRD WORLD COUNTRIES

	1981 crude Reserves (in 1000 barrels)	1981 Natural Gas Reserves (in 10^9 cu.ft)	1982 Annual Production (in 1000 barrels)
Mexico	48,083,772	75,352	1,003,630
Argentina	2,400,000	22,592	175,289
Brazil	1,475,911	2,118	94,761
Venezuela	20,149,703	48,210	675,166
Fucador	1,100,000	2,100	76,320
Algeria	6,941,680	123,550	352,225
Angola	1,427,714	1,561	46,680
Egypt	3,530,300	8,822	238,364
Gabon	463,488	38	53,613
Oibya	25,858,747	35,808	428,895
Nigeria	10,951,185	48,320	515,844
Tunesia	2,227,700	6,398	41,629
Iran	39,431,975	370,000	850,085
Iraq	33,700,000	25,250	357,744
Kuwait	67,680,000	33,000	304,962
Qatar	3,437,800	60,000	119,979
Saudi Arabia	158,499,603	113,393	2,291,044
Syria	1,392,622	3,147	62,913
Brunei	1,500,000	5,560	55,845
China	18,500,000	25,300	744,600
India	2,700,000	12,500	143,859
Indonesia	9,976,000	30,500	517,570
Malaysia	2,500,000	35,900	109,961
Australia	1,844,095	29,394	132,000
World Total	660,142,395	3,203,222	19,701,809

Source: World Oil v.196,No.3,p.170;v.195,No.3,p.36.

STRATEGIES FOR GEOPHYSICAL EXPLORATION OF MINERALS IN INDIA

B.B.Bhattacharya
Department of Applied Geophysics
Indian School of Mines
Dhanbad-826004

ABSTRACT

The future of exploration geophysics in developing countries including India appears to be enviably bright. The demand for basic natural resources will certainly grow in these regions, as the developing nations march forward to become developed nations, inspite of the availability of synthetic and substitute materials. The need for future mineral resources need not be over emphasized.

As the long term requirement of mineral resources will be considerable, even many of the low grade ore deposits, in course of time, will be economically viable. The efforts of exploration geophysicists in India would thus mostly be directed in the following ways:

a. Improvement of signal to noise ratio of geophysical methods and techniques to detect poorer and/or deeper ore bodies.

b. Increasing the depth of exploration of existing techniques, wherever possible, for the location of deeper ore bodies.

c. Extensive use of bore hole geophysical methods and application of geophysical techniques in underground workings wherever possible.

d. Wider application of microprocessor/computer for data acquisition analysis and interpretation.

e. Reinterpretation of geophysical data after exploratory drilling for minimising detailed drilling work.

f. Development of methods and techniques for possible prediction of composition of orebodies.

g. Application of airborne and ground geophysical methods in regional scale.

h. Development of new methods and techniques.

To achieve these objectives strengthening of geophysical departments of Indian Universities is equally important.

INTRODUCTION

The future of mining geophysics in India and developing countries appears enviably bright. Regardless of synthetic material as a substitute, the demand for basic resources will continue to expand and will accelerate as the populour underdeveloped nations begin to emerge. With the increasing demands, the incentives to find new resources to replace the mined materials would grow also. As new mines would be put into production, favourable areas of economic resources would be harder to find. It would then become necessary to take greater risks on poorer indications. The reliance obviously would tilt heavily on new technical methods or improved technical methods to find ore. No hypothesis or anomaly can ever rank as high as does visible ore inplace but such exploration guides are already becoming rare. This trend can be expected to become more acute.

The relative merits of any geophysical method in a given situation can be predicted by careful study of the expected "message-to-noise" ratio (Ward and Rogers, 1967). The word "message" is used to indicate the information sought whereas the word "signal" means summation of "message" and "noise". By estimating the physical property factor from the geologic knowledge of an area, one can predict the maximum anomaly to be expected for geometrical shaped targets (sphere etc.) at any depth of burial. However, one has to evaluate noise in order to determine the "message-to-noise" ratio. Noise encountered in geophysical surveys is of three types: instrument noise, disturbance field noise and terrain noise.

With any of the geophysical methods, the instrument noise can be, and has been, made lower than the threshhold of other noise sources. There are also ways and means to minimize, but never eliminate, the effects of disturbance field noise. The terrain noise for each method would include geologic noise and topographic noise plus location and sensor orientation. The terrain noise sources place practical but unknown limits on all of the methods. Thus each of the geophysical exploration method has a practical 'depth of exploration'. As terrain noise varies from location to location, the depth of exploration of any geophysical method remains practically unknown. Definition, evaluation and subsequent elimination of terrain noise is one of the greatest challenge facing the mining geophysicist today.

THE FUTURE TRENDS

The strategies for geophysical exploration in India and developing nations would possibly be directed in the following ways:

a. Improvement of signal to noise ratio of geophysical methods and techniques to detect poorer and/or deeper ore bodies.

b. Increasing the depth of exploration of existing techniques, whereever possible, for the location deeper ore bodies.

c. Extensive use of bore hole geophysical methods and application of geophysical techniques in underground techniques wherever possible.

d. Wide application of microprocessor/computer for data acquisition, analysis and interpretation.

e. Reinterpretation of geophysical data after exploratory drilling for minimising detailed drilling work.

f. Development of methods and techniques for possible prediction of composition of ore bodies.

g. Application of remote sensing including airborne geophysical surveys followed by ground geophysical methods for understanding the regional geological features.

h. Development of new methods and techniques.

REGIONAL AREA SELECTION AND PROSPECT IDENTIFICATION

The regional maps are a valuable data base for use by exploration agencies. They contribute directly and indirectly to many exploration programs. A systematic compilation of Remote Sensing, Aerial Geophysical, Deep Seismic Sounding (DSS), Regional gravity and Magnetic data is to be made for the preparation of base maps. Base maps at least in the scale of 1 : 100,000 or 1 : 250,000 covering standard map sheets should be prepared for the major mineralised belts.

Remote sensing is a relatively young scientific discipline. At the same time remote sensing also has roots in classical physics with ties to 19th century giants such as Maxwell, Fourier and Planck (Watson, 1985). The subject is limited to electromagnetic spectrum from 0.4 m to 25 cm. Three broad energy catagories are covered: solar reflected, thermal infra red and microwave.

The reflected solar energy region has been most intensely studied. Photointerpretation of images from individual spectral bands or colour composite remains the most widely used method of analysis. New instrumentation and digital processing, based on analysis of laboratory and field spectra, provide significant advances that are beginning to be applied to resource exploration. Colour composing techniques have been effectively used to detect the characteristic spectral reflectance features of iron oxides and hydroxyl bearing materials in satellite multispectral data for mapping areas of hydrothermal alteration. Airborne spectrometers can now detect individual spectral features of many minerals which are diagnostic of different stages of hydrothermal alterations. Evolution was from discrimination, based on empirical experience, to mineralogic identification and leading to quantification. Current developments also indicate new, promising extensions to vegetated terrains.

Advances in thermal infrared studies are due to development of thermal models that permit mapping of physical property variations and to tetection of spectral differences that provide important compositional information. Analysis techniques are still in their infancy and thermal satellite data remain appropriate only for regional investigations.

Microwave data have been acquired mainly with radar systems, which can provide very high resolution from space but use of textural and slope informations has had limited application. Long wavelength radiation has been shown to penetrate dry materials, and this may be applicable in extremely arid regions.

Prospect identification refers to the use of reconnaissance geophysical methods to reduce area of interest and to select areas for more detailed work. Primarily the low level detailed aerial geophysical surveys using magnetic, radiometrics and electromagnetic methods would be immensely useful. Magnetic airborne survey results are presented as contours or profiles of residual total magnetic intensity after subtraction of the International Geomagnetic Reference Field (IGRF).Radiometric surveys normally use multichannel gamma-ray spectrometers. The results are presented as profiles or contours of count rates for total count (TC), potassium, uranium and thorium. If the system is calibrated, the results may also be presented as parts per million equivalent potassium, uranium or thorium.

Historically total-count radiometric surveys, and, more recently, gamma-ray spectrometry have been used in direct prospecting for uranium. In addition to direct detection of anomalies due to mineralization, radiometric surveys can also be used to refine geologic mapping and thus aid target selection. The simultaneous display of gamma-ray spectrometer data from three sources (potassium, uranium, and thorium decay series) is difficult to achieve with conventional black and white contours or multiparameter profiles. The application of colour displays and image processing techniques can significantly enhance these data. It is possible to view simultaneously all three channels of data in a plan view and thus gain a greater appreciation of regional patterns, structures etc (Richards et al.,1983).

In EM airborne surveys frequency and time domain techniques are in use. In recent years, however, time domain is being used increasingly.

TARGET SELECTION AND DRILLING

Detailed prospecting leading to selection of drilling targets usually involves the use of one or more ground based geophysical methods, often as a follow up to an airborne survey. Ground magnetics, gravity, electrical (self-potential, resistivity, induced-polarization (IP) and electromagnetic (both frequency and time domain) methods are all used to discriminate between possible targets and to improve target definition for drilling. In exploration of sulphide mineralization at shallower depths (less than 150 m) electrical and electromagnetic methods are commonly used. In spite of deep and often conductive cover, these methods have been effective and, of course, they are being constantly improved. In addition to the conductive cover, the effect of conductive host on the response of the target for electrical methodfs is a topic of interest to the present day research workers.

The next stage of exploration, drilling, is frequently accompanied by drill-hole geophysical surveys and geophysical logging for the delineation of the ore body and optimizing detailed drilling work. The drill-hole geophysical surveys are used to search for "near misses". Almost all the electrical and electromagnetic methods are used for drill hole geophysical surveys. Magnetic methods are also in constant use. But mise-a-masse (MALM) method is veryextensively used. In this method the ore body struck in the bore-hole is energized by placing an electrode at it. The potential measurements are made either at the surface or in the adjoining Bore holes to determine size and attitude of body struck in the bore-hole.

In uranium and coal exploration, open hole drilling followed by a suite of geophysical logs is the basis of most exploration programs. The primary tools are natural gamma, density and caliper logs. These methods are widely used and are well documented in the literature. The traditional logging tools (e.g.,SP, resistance, natural gamma, neutron, etc) which were primarily developed for petroleum exploration in sedimentary rocks are less useful in mineral exploration except in coal exploration. Alternative tools (e.g., magnetic susceptibility, induced polarization, etc.) do exist and are used from time to time.

EVALUATION AND MINE DEVELOPMENT

A great deal of orebody evaluation consists of drilling and logging (including geophysical logging). This is particularly so in coal and uranium evaluation programs where geophysical logs can given reliable quantitative estimates of ore grade, impurities etc. Interpolation between drill holes and guides to drill hole location can be obtained from surface and drill-hole geophysical methods.

SOME RECENT METHODS

Apart from the various geophysical methods mentioned above which are well established in exploration programs, a few new methods have been attempted which are not yet used in a routine fashion. The methods are still in nascent stage but appear to be promising. These methods are (i) Contact Polarization Curve (CPC), and (2) Nonlinear Induced Polarization (NIP).

Contact Polarization Curve (CPC) Method

The CPC method has been attempted in USSR (Ryss, 1972). It is based upon the cathodic and anodic reactions that take place when an electric current passes through an electrolytic cell. If two metallic or, more generally, electronically conducting electrodes are placed inan electrolyte each assumes a definite potential with respect to the electrolyte. It is necessary to give such electrodes an external electric tension, called electrolytic polarization, to obtain electrochemical equilibrium. An example would be a cell consisting of iron and carbon electrodes in a solution of zinc sulphate. As the external tension applied to this cell is increased there is first only a very small, almost vanishingly small, flow of current when the tension is low. This is the so called nonfaradic current. Above the certain value of the applied tension, however, a strong flow of current begins. The critical tension at which this so called farradic current starts to flow depends upon the material of the electrode an also whether the electrode is a cathode or anode. It can be determined from the intersection of the voltage current curve with voltage axis.

Many minerals, e.g., pyrite, chalcopyrite, galena etc are electronic conductors and show the above phenomena when used as cathodes or anodes in an electrolytic cell.

Nonlinear Induced Polarization (NIP) Method

It is a geophysical method based on those characteristics of the secondary electric field which do not obey a linear relationship at the interface between electronically conducting and ionically conducting media. In otherwords, the relationship between the current and voltage is nonlinear though the existing theory of IP assumes a linear relationship between e.m.f. and current density for IP. However, in laboratory tests deviations from linearity for these two quantities are observed for current densities exceeding some critical values (Komarov et al, 1972). For low current density discrepancy between the cathodic and anodic polarization of electronic conductor still exists. This difference is utilized in the non-linear IP method. The bore hole version of the nonlinear IP method has also been used.

NIP measurements can be carried out with standard IP equipment, both on the surface and in boreholes. It is claimed that the method can be used for practical discrimination of the IP anomalies, in order to discriminate sulphide ore bodies from pyritic and graphitic bodies.

CONCLUSIONS

The geophysical methods have made real contributions to economic discoveries or significant extensions of known reserves. Geophysics has contributed to regional area selection, prospect identification, target selection, drilling, ore body evaluation and mine planning and will continue to do so by applying the existing methods with improved measurement and data processing techniques and by developing new methodologies for measurements and data processing. For successful and cheaper exploration the application of geophysics will be much wider in future than it is today.

REFERENCES

Komarov,V.A., Ioffe,L.M.,and Smirnov,A.A.,1972. The principles of nonlinear IP (NIP) method : Methods and Technology in Prospecting, collection of works, No.81 (in russian).

Richards,D.J., Simon,I., O'Sullivan,K.,And Hogan,A., 1983. An image processing system applied to airborne radiometric data: Abstracts, Austral.Soc.Exp.Geophy. Biennial Conference, Brisbane.

Ryss,Yu.S., 1973. Search of ore bodies by Contact Polarization Curve Method: Leningrad, Nedra, 167 p (in russian).

Ward,S.H., and Rogers, G.R.,1967. Introduction: Mining Geophysics, Vol.II, Society of Exploration Geophysicists,p.3-8.

Watson, Ken, 1985. Remote Sensing - a geophysical perspective : Geophysics, vol.50, No.12, p.2595-2610.

STRATEGY FOR EXPLORATION OF GONDWANAS IN PENINSULAR INDIA

R K Verma
Department of Applied Geophysics
Indian School of Mines
Dhanbad, India.

INTRODUCTION

Gondwanas are the most important geological formation of Peninsular India from the point of view of coal resources. They are preserved in sedimentary basins or troughs which have been formed during Permo-Carboniferous to Mesozoic times. The areas where these formations have been found so far are in Godavari Valley, Mahanadi Valley, Satpura basin, Damodar Valley, Rajmahal hills and their extension into Bengal basin.

Although Gondwana formations exposed near the surface have already been mapped, there are large parts of Peninsular India where Gondwana formations underlie the younger formations such as Deccan trap, Rajmahal trap and Tertiary sediments. The areas where the Gondwanas are at present exposed and their probable extension into adjoining areas is shown in Figure 1. A detailed study and analysis of gravity field can enable us to map these areas and to determine the extension of Gondwanas underneath the younger formations. Analysis of gravity field in different basins of India, their basement structure and probable thickness of Gondwanas is discussed in this paper.

GODAVARI VALLEY

Bouguer anomaly map of Godavari Valley and the surrounding regions (after Qureshy et al., 1968) is shown in Figure 2. The gravity field shows a well-defined minimum of about -40 mgal amplitude in the central part of the valley. The low is attributable to the low density (about 2.4 gm/cm^3) Gondwana formations filling the trough. The gravity contours are elongated in NW-SW direction, nearly parallel to the strike of Godavari valley. In coastal areas, south of Bhadrachalam, the contour assume a NE-SW strike direction, parallel to the Eastern Ghats strike. To the northwest of Godavari valley (beyond Chanda), two well defined lows are found. One near Nagpur and the other near Sohagpur. This suggests that the Gondwanas may extend in northwesterly direction upto Sohagpur and even beyond.

The Basement structure analysed along two profile, AA' and BB' (see Fig.2) is shown in Figure 3 (after Qureshy et al.,1968). The figure shows that the Godavari valley has a double walled graben structure, the northern side dipping more steeply as compared to the southern side. Overall structure of the valley appears to be that of a rift. Maximum thickness of sediments in the valley is estimated to be of the order of 4.2 km.

MAHANADI VALLEY

Detailed gravity anomaly map of Mahanadi valley has been prepared by Subba Rao (1977), as shown in Figure 4. It may be seen that a major low extends from near Hingir in the south to near Umaria in the north via Chokan and Shahdol, in northwesterly direction. The low correlates well with the Lower Gondwana formations of Mahanadi valley. The gravity lows interpreted in terms of basement structure along two profile AA' and EE' (location shown in Figure 4) are shown in Figure 5.

The Mahanadi valley appears to consist of several grabens which are separated from each other by crystalline metamokrphics. The Gondwanas appear to terminate suddenly north of Barach and Weidhan along a major fault, which separates the Vindhyas from the Gondwanas. A detailed analysis of gravity field in all the basins is necessary in order to ascertain the overall potential of coal bearing formations in the area.

SATPURA BASIN

As shown in Figure 1, the Gondwana formations belonging to Satpura basin extend in ENE-WSW direction along the Narmada-Son line. The basin appears to merge with Mahanadi basin further north east. A detailed Bouguer anomaly map of part of the entire basin (after Verma and Banerjee, unpublished) is shown in Figure 6. The figure shows that the Gondwanas forming a part of this basin extend between latitude 22° to 23°N and longitude 77.5° to 79.5°E. The gravity field suggests that the Gondwanas may extend towards northwest and west of Sohagpur underneath the alluvium and/or Deccan traps. This has to be proved by drilling. An interpreted profile AA' is shown in Figure 7 (location shown in Figure 6).

The figure shows that the Sohagpur coalfield forms a part of a very major Satpura basin, extending over a distance of nearly 160 km in N-S direction and over 200 km in east-west direction. Besides the Satpura basin, the structure of Kamthi and Wardha basins is also shown in the figure.

It may be mentioned here that the Gondwanas appear to extend to an appreciable extent underneath Deccan traps towards Ujjain. This has been confirmed recently by DSS profile from Ujjain to Mahan as shown by Kaila et al. (1984)

DAMODAR VALLEY GONDWANA GRABENS

Raniganj, Jharia, Bokaro, Karanpura, Ramgarh and Daltonganj form an east-west trending belt of Gondwana grabens extending from longitude 83.5°E to 87.5°E and latitude 23.5°N to 24.2°N. A detailed study of gravity field for these basins has been carried out by Verma and Ghosh (1974, 1977) and Verma et al. (1976, 1979, 1983). Detailed analysis of gravity field has been carried out in terms of three-dimensional models for Jharia and Raniganj coalfields. A model for the basement underlying Jharia coalfield base on gravity data is shown in Figure 8. An interpreted profile AA', crossing the basin in NW-SE direction is shown in Figure 9. The geological section also shows the Talchir, Barakar and Raniganj formations with depth along the profile. The location of various coal seams is projected from their surface exposures with depth. Interpretation of several similar sections across Raniganj and Jharia coalfields has been discussed by Verma et al. (1979, 1980).

RAJMAHAL HILLS AND BENGAL BASIN

Gondwana formations occur over a very wide area along the eastern margin of the Peninsular shield as shown by several boreholes drilled in Bengal basin. They underlie mostly the Tertiary formations of the Bengal basin, extending from Port Canning (22°N, 88°.7E) in the south to Kishanganj (26.2°N, 88°E) in Purnea basin as shown in Figure 10 (after Choudhury and Dutta, 1973).

In addition to above, Gondwanas have been found to be underlying Rajmahal traps as shown from analysis of gravity field (Ashraf, 1978). The estimated thickness of Gondwanas in Bengal basin is 2.0 to 2.5 km. An interpreted gravity profile extending from the eastern margin of Raniganj coalfield to Calcutta is shown in Figure 11 (after Choudhury and Dutta, 1973). The figure shows that a considerable thickness of Gondwana formation underlies the younger formation in Bengal basin. There is need for accurately assessing the potential of Gondwanas in different parts of the Bengal basin. This can be done by analysing the gravity field, using seismic data (wherever available), combined with few borehole observations.

SUMMARY AND CONCLUSIONS

Since our coal reserves are found mostly in Lower and Upper Gondwana formations, an understanding of the Gondwanas, their areal extent and depth in various parts of Peninsular India is very essential. What we need basically is information regarding the thickness of Gondwanas, and to project it with depth. This can be done through detailed gravity surveys and analysing the gravityfield in terms of two dimensional or three dimensional models. Wherever possible seismic data, DSS data and borehole information should be used to test our models.

It appears that coal reserves may be present over large areas underlain by Deccan traps, Rajmahal traps, alluvial areas in Satpura basin and underneath Tertiary and Mesozoic formations in Bengal basin. There is need for undertaking detailed geophysical investivations in these areas.

REFERENCES

Ashraf,M.H., 1978. Results of Gravity Surveys over Some Parts of Eastern India. Ph.D. ttthesis, Indian School of Mines, Dhanbad (unpublished).

Choudhury, S.K. and Datta, A.N., 1973. Bouguer Gravity and Geological Evaluation in Western part of Bengal Basin and Adjoining Areas. Geophysics, 38,p.691.

Kaila, K.L., Reddy,P.R., Dixit, M.M. and Rao, Koteswar, 1985. Crustal structure across Narmada-Son Lineament, Central India from deep Seismic Sounding. Jour.Geol.Soc.India, 26,pp.465-480.

Qureshy,M.N., Krishna Brahman, N., Garde,S.C. and Mathur,B.K., 1968. Gravity Anomalies and the Godavari Rift, India. Geol. Soc.of Am.Bull., 79, 1221-1230.

Subba Rao,D.V., 1977. Gravity Studies over Mahanadi valley and South Rewa Gondwana Basins of India. Ph.D. thesis, Andhra University, Waltair.

Verma,R.K. and Ghosh,D., 1974. Gravity Survey over Jharia Coalfield, India. Geophy.Res.Bull.,12, 49-59.

Verma,R.K.,Majumdar,R., Ghosh,D.,Ghosh,A and Gupta,N.C.,1976. Results of Gravity Survey over Raniganj Coalfield, India. Geophy. Prosp.,24, 19-30.

Verma,R.K. and Ghosh, D.,1977. Gravity field, Structure and Tectonics of Some Gondwana Basins of Damodar Valley, India in Geophysical case histories of India, v.1, 97-112.

Verma,R.K., Bhuin,N.C. and Mukhopadhyay,M., 1979. Geology, Structure and Tectonics of Jharia Coalfield, India - A Three-dimensional Model. Geoexploration, 17, 305-324.

Verma,R.K.,Bandopadhyay,T.K., and Mukhopadhyay,M.,1980. Gravity field, Structure and Tectonics of the Raniganj Coalfield - Three Dimensional model. J.Geol.Soc.India, 21, 117-127.

Verma,R.K.,Mukhopadhyay,M. and Ashraf,M.H.;1983. Gravity field and Tectonics in Karanpura-Daltonganj Coalfield, Eastern India. J.Geol.Soc.India.

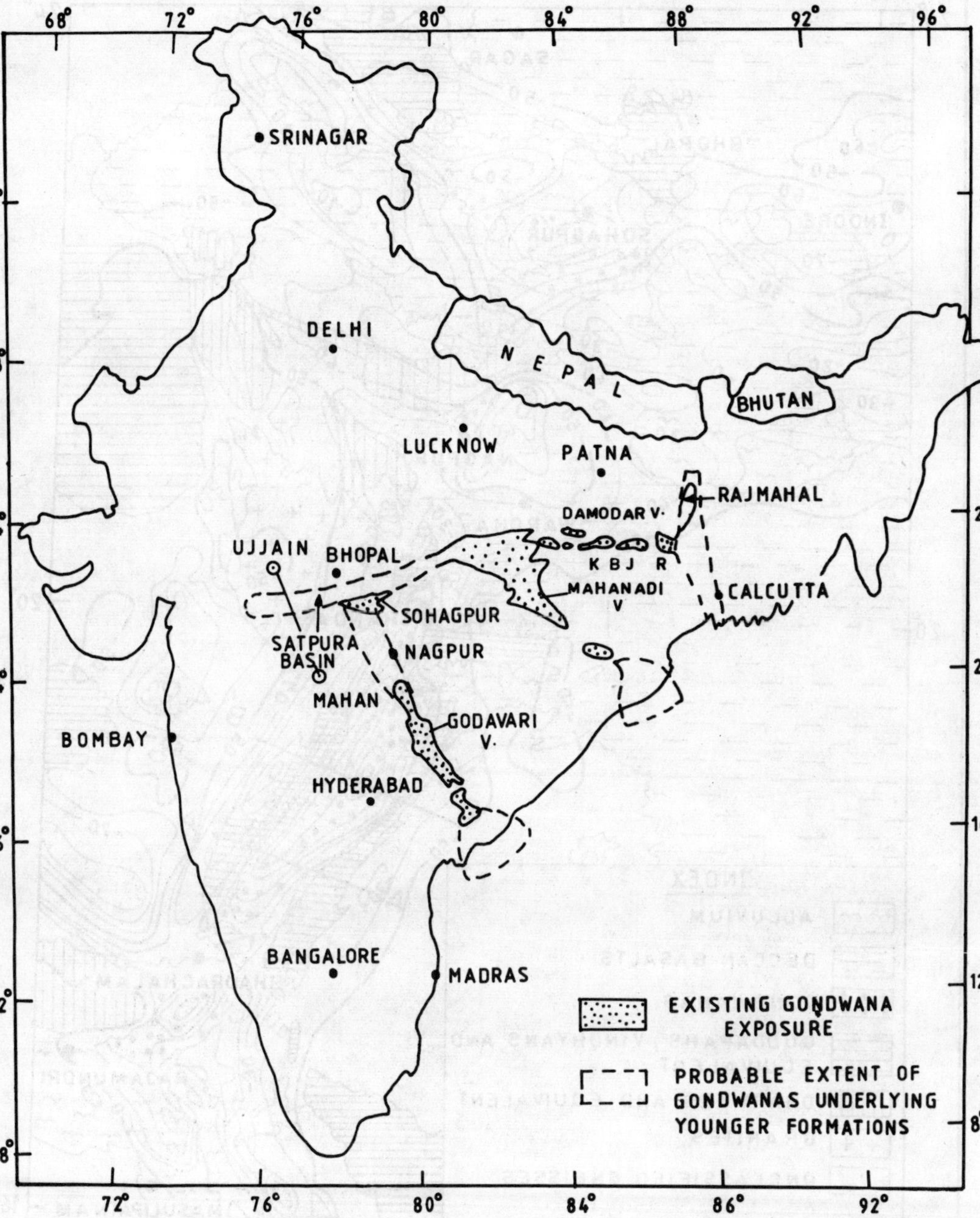

ABBREVIATIONS: B - BOKARO, J - JHARIA, K - KARANPURA, R - RANIGANJ

FIG. 1

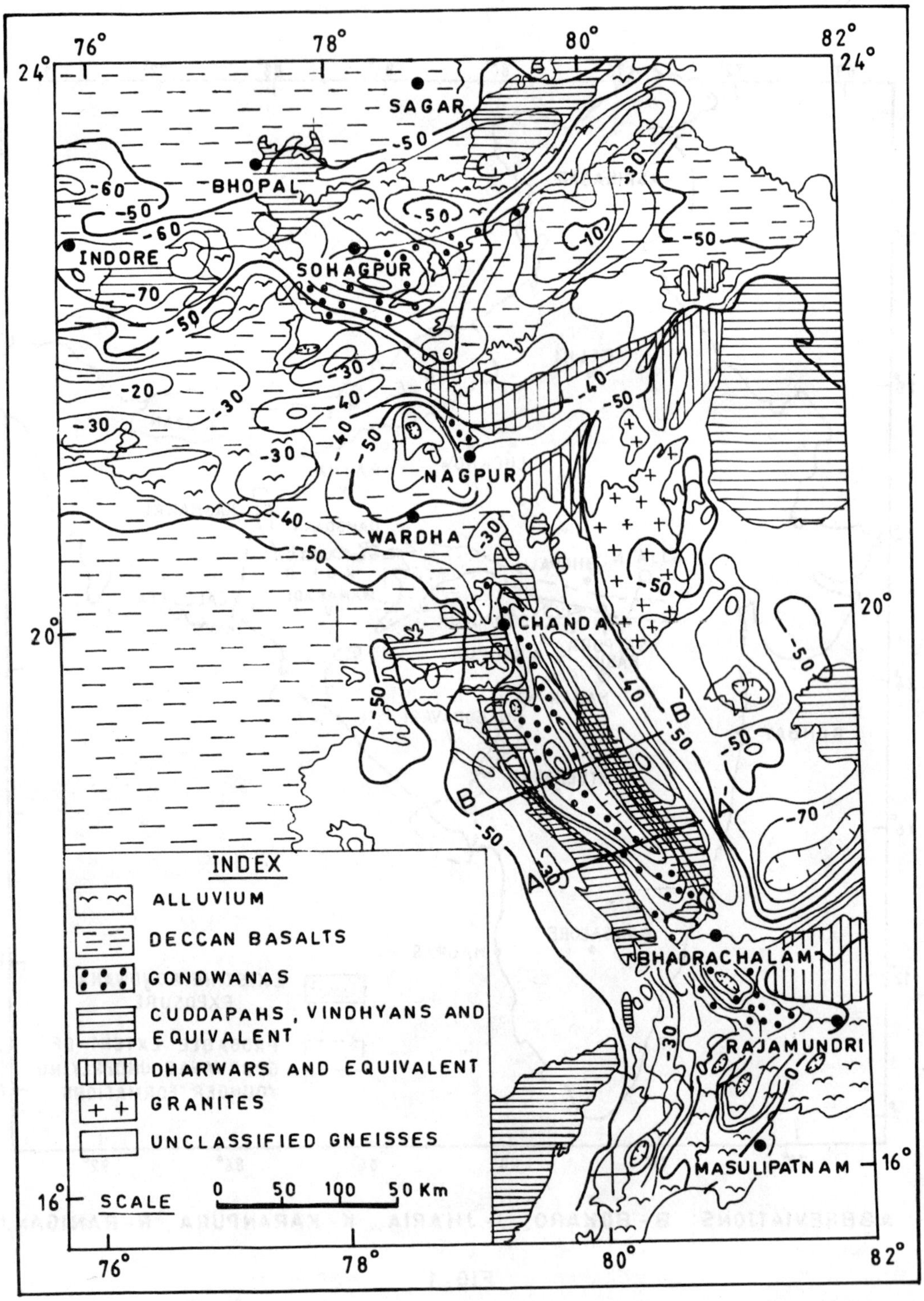

INDEX
ALLUVIUM
DECCAN BASALTS
GONDWANAS
CUDDAPAHS, VINDHYANS AND EQUIVALENT
DHARWARS AND EQUIVALENT
GRANITES
UNCLASSIFIED GNEISSES
SCALE
0 50 100 50 Km
SAGAR
BHOPAL
INDORE
SOHAGPUR
NAGPUR
WARDHA
CHANDA
BHADRACHALAM
RAJAMUNDRI
MASULIPATNAM
76°
78°
80°
82°
24°
20°
16°

FIG. 2

GODAVARI VALLEY

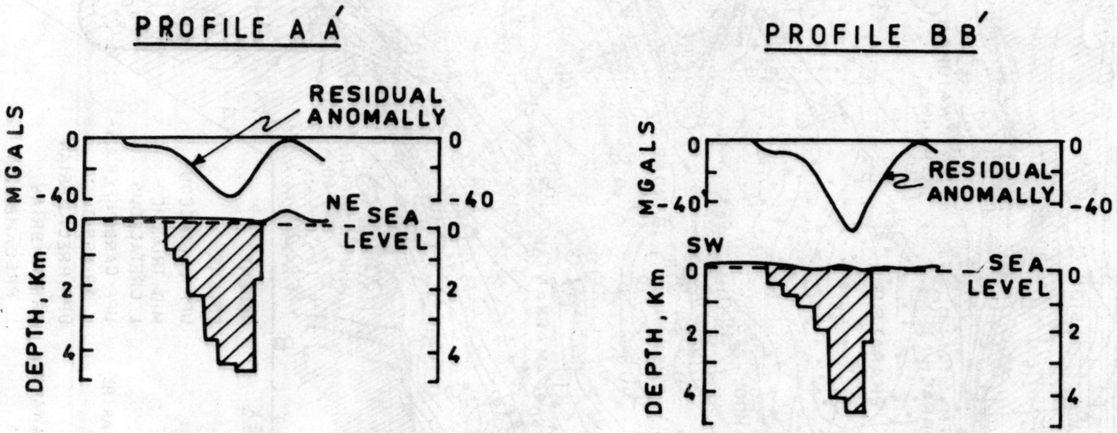

FIG. 3

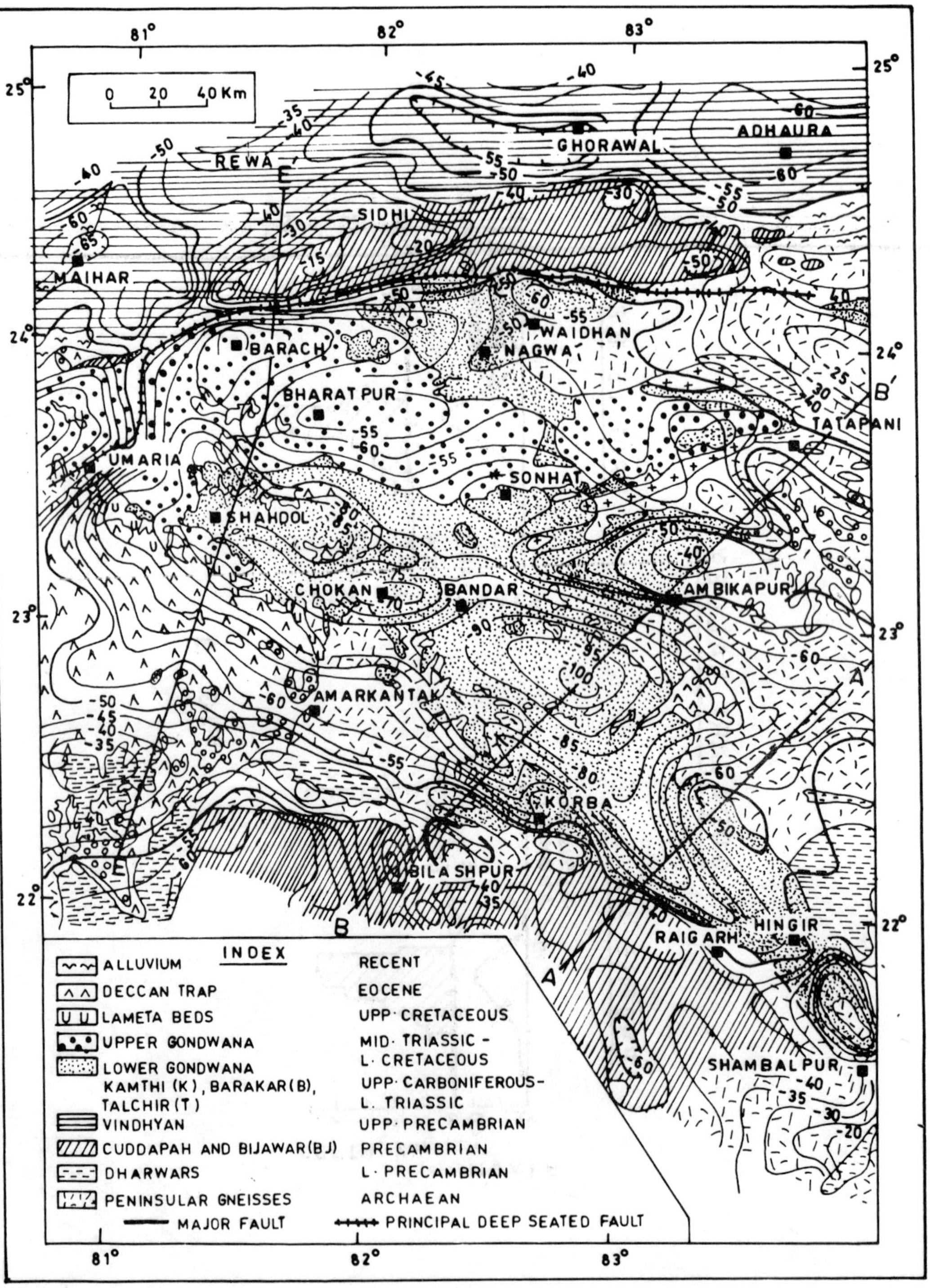
0 20 40 Km
REWA
SIDHI
GHORAWAL
ADHAURA
MAIHAR
BARACH
WAIDHAN
NAGWA
BHARATPUR
TATAPANI
UMARIA
SONHAT
SHAHDOL
CHOKAN
BANDAR
AMBIKAPUR
AMARKANTAK
KORBA
BILASHPUR
RAIGARH
HINGIR
SHAMBALPUR
INDEX
ALLUVIUM RECENT
DECCAN TRAP EOCENE
LAMETA BEDS UPP. CRETACEOUS
UPPER GONDWANA MID. TRIASSIC - L. CRETACEOUS
LOWER GONDWANA KAMTHI (K), BARAKAR (B), TALCHIR (T) UPP. CARBONIFEROUS - L. TRIASSIC
VINDHYAN UPP. PRECAMBRIAN
CUDDAPAH AND BIJAWAR (BJ) PRECAMBRIAN
DHARWARS L. PRECAMBRIAN
PENINSULAR GNEISSES ARCHAEAN
MAJOR FAULT
PRINCIPAL DEEP SEATED FAULT

FIG. 4

PROFILES ACROSS MAHANADI VALLEY

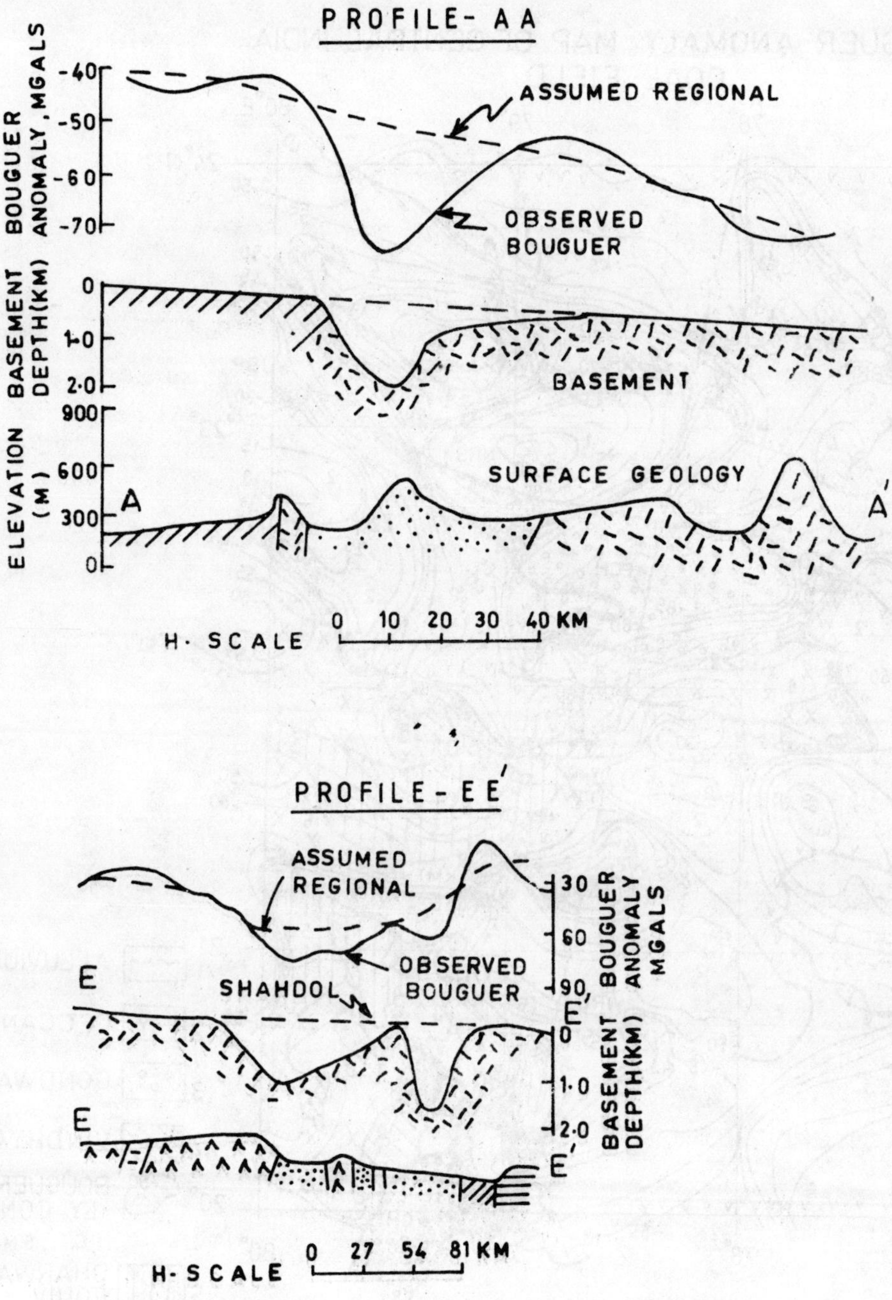

FIG. 5

BOUGUER ANOMALY MAP OF CENTRAL INDIA COAL FIELD

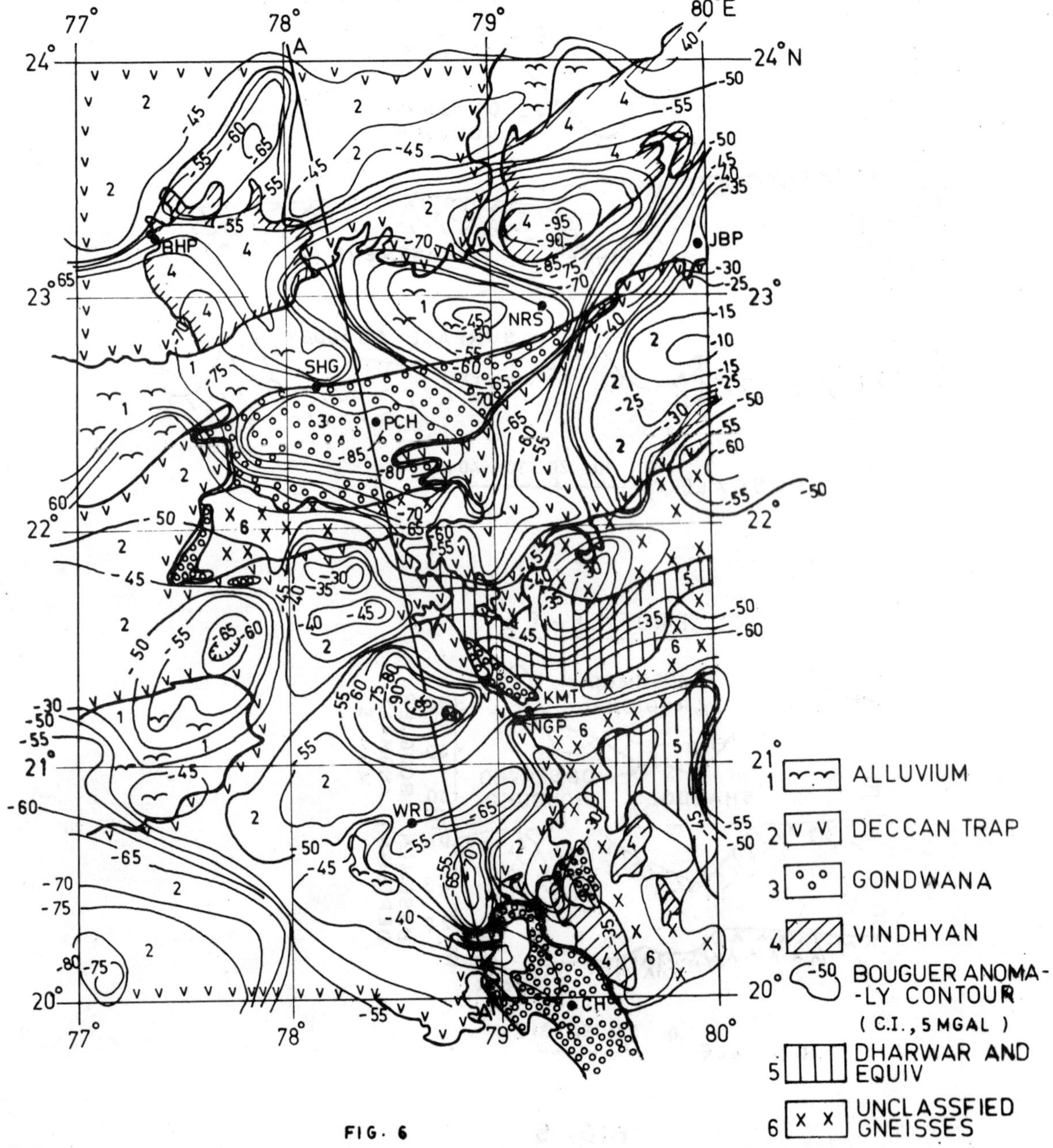

FIG. 6

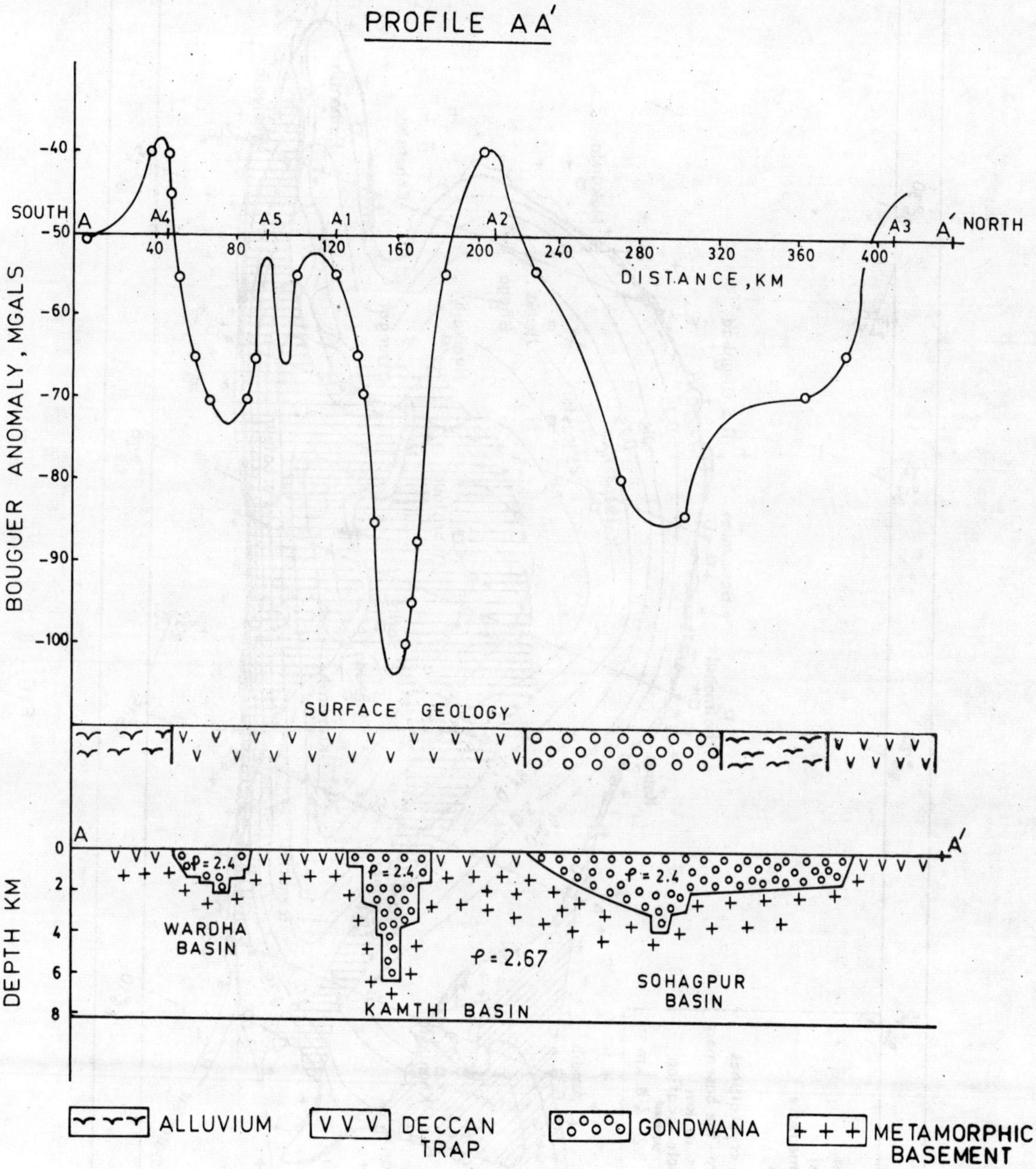
PROFILE AA′
BOUGUER ANOMALY, MGALS
-40
-50
-60
-70
-80
-90
-100
SOUTH
NORTH
A
A4
A5
A1
A2
A3
A′
40
80
120
160
200
240
280
320
360
400
DISTANCE, KM
SURFACE GEOLOGY
DEPTH KM
0
2
4
6
8
ρ = 2.4
ρ = 2.4
ρ = 2.4
ρ = 2.67
WARDHA BASIN
KAMTHI BASIN
SOHAGPUR BASIN
ALLUVIUM
DECCAN TRAP
GONDWANA
METAMORPHIC BASEMENT

FIG · 7

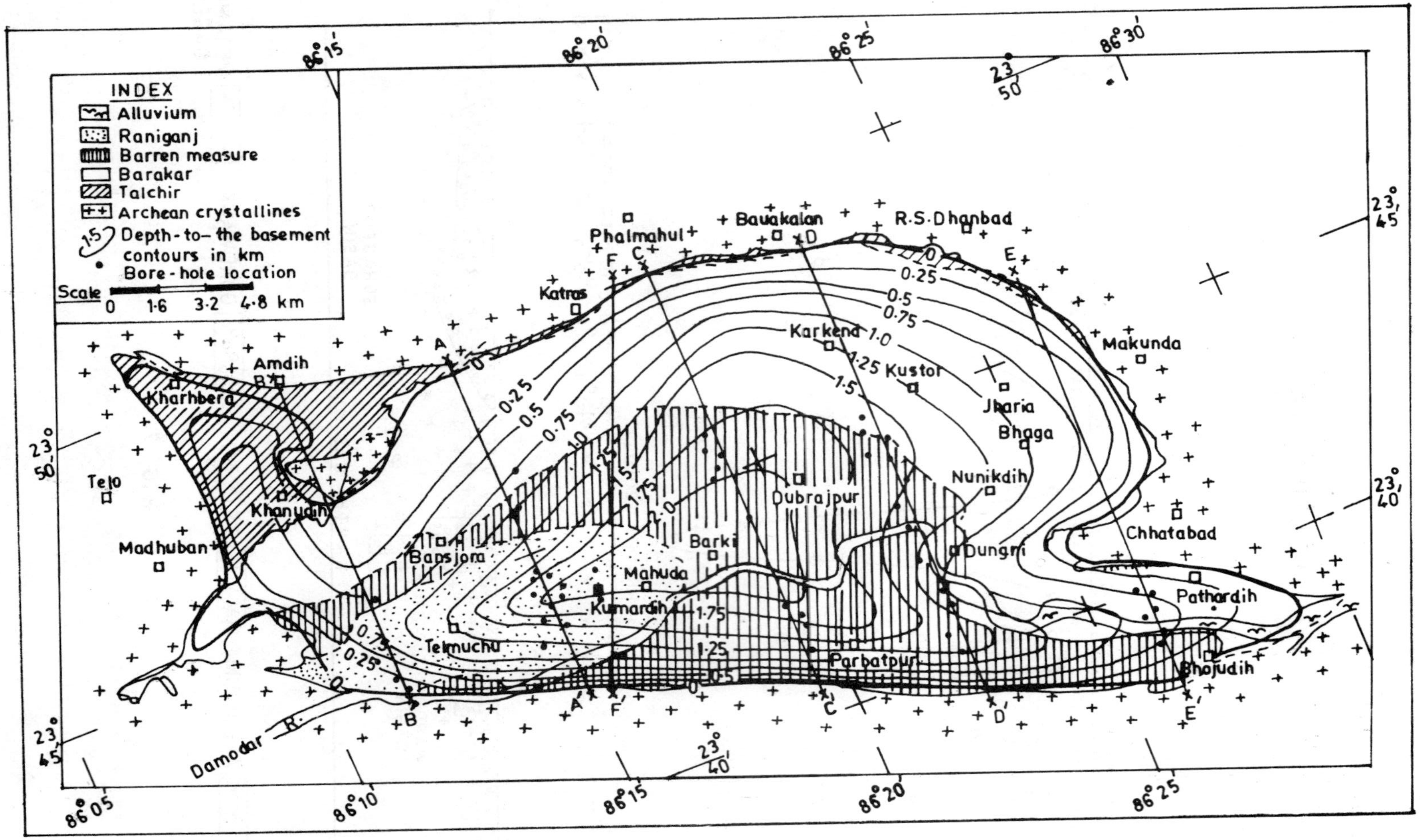
INDEX
Alluvium
Raniganj
Barren measure
Barakar
Talchir
Archean crystallines
Depth-to-the basement contours in km
Bore-hole location
Scale 0 1·6 3·2 4·8 km
Amdih
Kharhbera
Telo
Madhuban
Khanudih
Katras
Phalmahul
Bauakalan
R.S. Dhanbad
Makunda
Karkend
Kustor
Jharia
Bhaga
Nunikdih
Dubrajpur
Barki
Bansjora
Mahuda
Kumardih
Telmuchu
Dungri
Chhatabad
Pathardih
Parbatpur
Bhojudih
Damodar R.

FIG . 8

PROFILE AA', JHARIA COALFIELD BASED ON GRAVITY AND BOREHOLE DATA

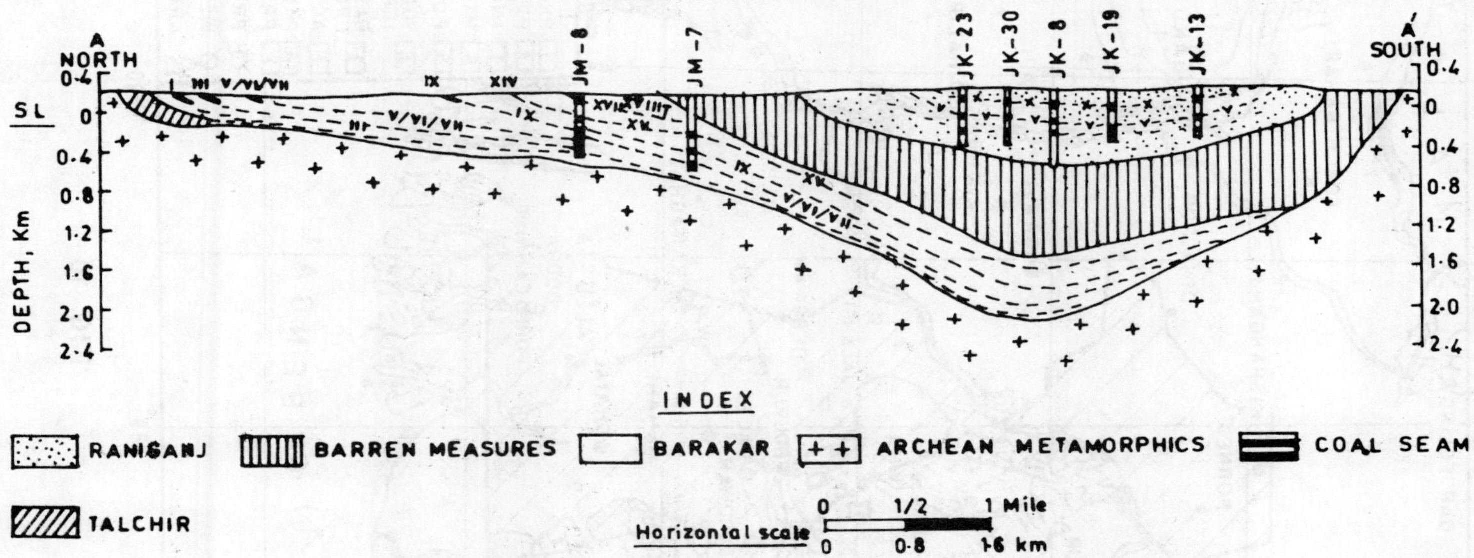

FIG. 9

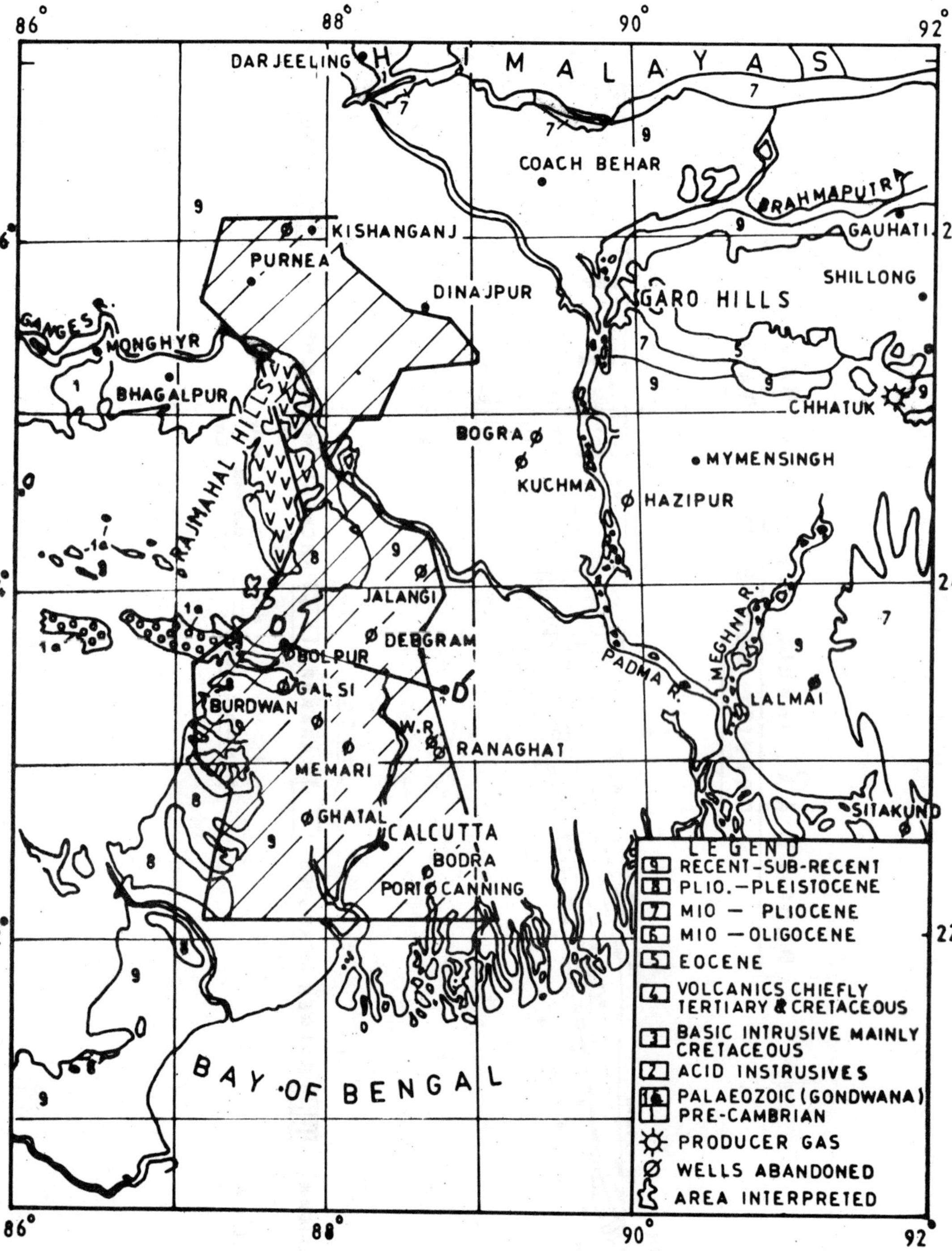

86°
88°
90°
92°
26°
24°
22°
DARJEELING
HIMALAYAS
COACH BEHAR
BRAHMAPUTRA
GAUHATI
SHILLONG
GARO HILLS
CHHATUK
KISHANGANJ
PURNEA
DINAJPUR
GANGES R.
MONGHYR
BHAGALPUR
RAJMAHAL HILLS
BOGRA
KUCHMA
MYMENSINGH
HAZIPUR
JALANGI
DEBGRAM
BOLPUR
GALSI
BURDWAN
W.R
RANAGHAT
MEMARI
GHATAL
CALCUTTA
BODRA
PORT CANNING
PADMA R.
MEGHNA R.
LALMAI
SITAKUND
BAY OF BENGAL
LEGEND
9 RECENT-SUB-RECENT
8 PLIO.-PLEISTOCENE
7 MIO - PLIOCENE
6 MIO - OLIGOCENE
5 EOCENE
4 VOLCANICS CHIEFLY TERTIARY & CRETACEOUS
3 BASIC INTRUSIVE MAINLY CRETACEOUS
2 ACID INSTRUSIVES
1a PALAEOZOIC (GONDWANA)
1 PRE-CAMBRIAN
PRODUCER GAS
WELLS ABANDONED
AREA INTERPRETED

FIG. 10

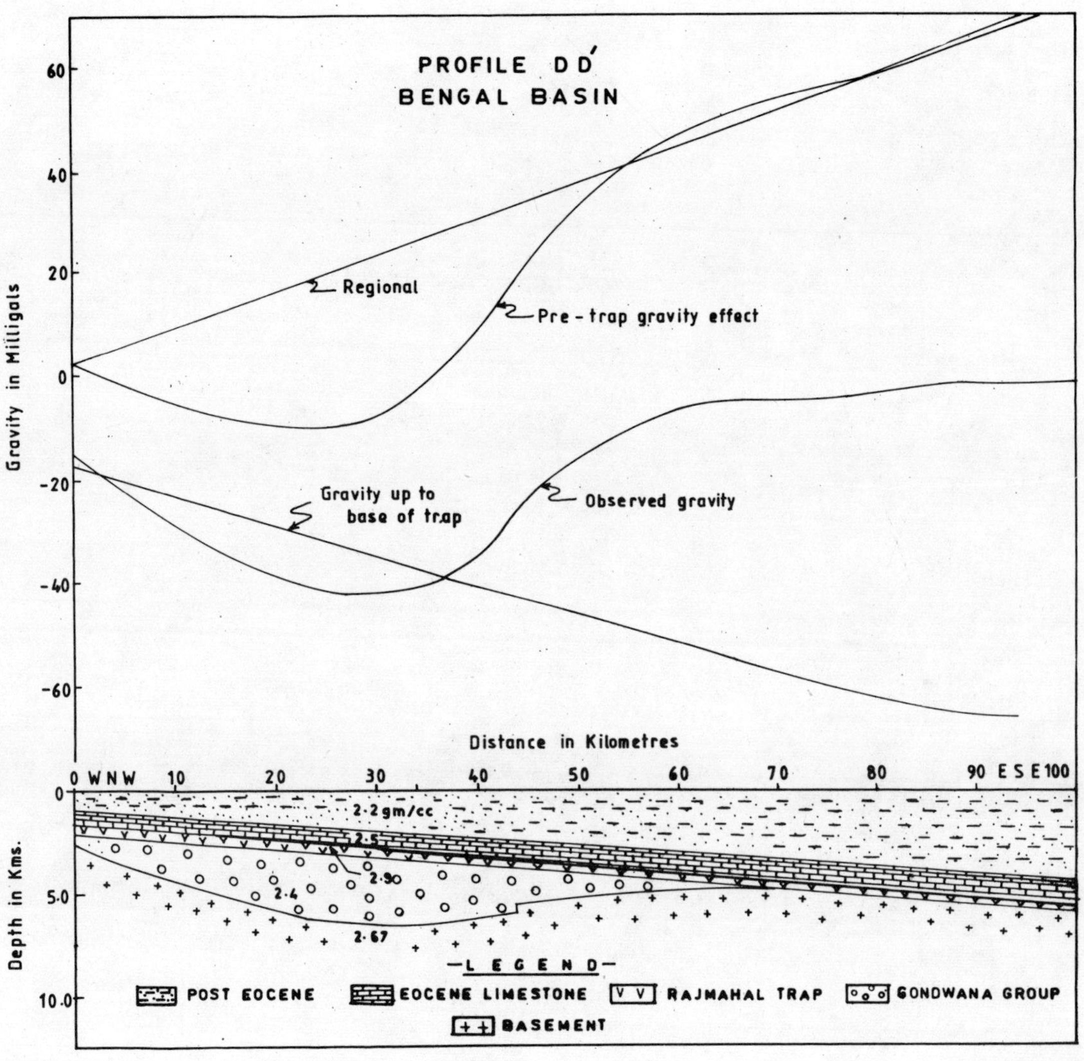

PROFILE DD'
BENGAL BASIN
Gravity in Milligals
60
40
20
0
-20
-40
-60
Regional
Pre-trap gravity effect
Observed gravity
Gravity up to base of trap
Distance in Kilometres
0 WNW 10 20 30 40 50 60 70 80 90 ESE 100
Depth in Kms.
0
5.0
10.0
2·2 gm/cc
2·5
2·90
2·4
2·67
— L E G E N D —
POST EOCENE
EOCENE LIMESTONE
RAJMAHAL TRAP
GONDWANA GROUP
BASEMENT

FIG. 11

PROBLEMS AND PROSPECTS OF SMALL-SCALE METAL MINING IN THE METALLOGENIC DOMAINS OF THE HIMALAYAS

K.L. Rai
A.Madhusudana Rao
Indian School of Mines, Dhanbad

ABSTRACT

In their geotectonic setting the Himalayas are comparable to the mountain - chains of Appalachians in USA, Urals and Caucasus in USSR and Alps in Europe, all of which are endowed with rich metallization. Unfortunately, however, in spite of fairly detailed exploration in its various metallogenic domains during the last 30-35 years, the ore search has proved largely futile and disappointing so far.

The patterns of distribution of ore mineralisations in the Himalayas in terms of time and space have been briefly reviewed. In this light, the promising metallogenic domains and deposits that have proved amenable to small-scale mining, particularly in the sub-Himalayan zone, have been identified.

The problems of exploration on various scales and of the exploitation of ores in such deposits have been discussed.

In conclusion, a strategy that calls for basic philosophical change in our attitude and concepts regarding the techno-economic feasibility of mining in such difficult yet strategically significant part of the country has been proposed. In the context of the dismal failure of our existing strategy, the need for immediate change in it has been emphasized.

INTRODUCTION

The Himalayas constitute possibly the youngest yet a potential metallogenic province of the Indian sub-continent. Large-scale exploration activity in its various parts particularly during the last three decades has yielded a wealth of valuable data on numerous mineral occurrences, many of which have shown the promise of workability as paramarginal, if not economic, resources under the present-day techno-economic norms.

The paper aims at highlighting the potentialities of the great Himalayan mountain chain with special reference to the workable metallic mineral resources of its promising metallogenic domains and the possibility of exploiting some of them under a planned geological strategy for small-scale operations. Impending need for the exploitation of these resources has to be viewed in the context of the desirable socio-economic development of this vast 'terra incognita' which extends over 1500 km stretch of the country's strategic border in the north and east.

METALLOGENIC DOMAINS AND THEIR DISTRIBUTION PATTERN

The metallogenic domains identified so far exhibit delineable patterns of their distribution in time and space (Table 1).

Geological distribution

Numerous occurrences of metallic minerals, many of them promising ones from points of view of small-scale mining are known to be localised in the Himalayan formations ranging from Precambrian to Neogene in age. The Precambrians alone account for more than 80% of the workable ore resources proved so far. Only one of them namely the Bhotang deposit at Rangpo in Sikkim Himalaya has, however, been found economically viable and is under active exploitation at present.

As compared to the Precambrians, the Palaeoxoics (Caledenian) mark an era of subdued metallogenic activity of Cu, Pb, Zn, and Sb mineralization often in association with the arenaceous-carbonate suite of rocks in the Western and Central Himalayas. By contrast, the Meso-Cainozoics exhibit virtual absence of any potential mineralization almost throughout the Himalayas, this being a feature conspicuously opposed to the global pattern which is generally marked with rich metallizations.

Distribution in space:

Conforming to the prevailing geotectonic and physiographic features, Krishnan (1956) Gansser, 1964, and various other geoscientists of the Geological Survey of India have identified six linear metallogenic zones in the Himalayas (Fig. 1). These zones, placed in proper order of their settings from north to south and their mineral potentialities may be briefly stated as follows:

1) Trans-Himalayas comprising granitoid massifs with possible presence of porphyry-type Sn, W, Bi, Mo and REE mineralization.

2) Ophiolite zone of basic-ulterabasic suite of rocks with Cr, Ni, Cu, Pt and Fe mineralizations.

3) Tethys geosyncline zone of marine sediments and granitoids, with minor occurrences of base metals and chromite.

4) Central Himalayan Zone (of about 65 km width) dominated by granitoids containing W,Bi, Mo, Sb and REE mineralizations.

5) Lesser Jo,a;auam zpme (of about 80 km width) having a promising volcano-sedimentary suite of rocks containing sporadic Pb, Zn, Zn, Cu, Sb and possibly Hg mineralizations.

6) Outer Himalayan/Siwalik zone (of about 50 km width) containing sporadic occurrences of copper in the Red-beds of Tertiary sediments and Uranium often with rock phosphates inthe Siwalik sediments.

The unfossiliferous Lesser Himalayan zone south of the Main Central Thrust constitutes the most promising metallogenic domain particularly in its Precambrian segments. These segments have been regarded as the probable outliers of northern prolongations of the Indian Shield elements in Himalayan foot-hills by many geoscientists (Wadia 1938 & 1956, Gansser 1964). Studies by Gansser (1964), Dar (1968) and others have shown that the maximum concentration of the base-metal deposits occur along or in the vicinity of the northern prolongations of the following shield elements (Fig.2):

a) Aravalli strike continuous (with Delhi - Haridwar ridge) in Kulu-Simla-Garhwal-Kumaun regions.

b) Bifurcated extension of Aravalli Strike (or Lahore-Sargora Ridge) in Hissar-Kirana sector of the Kashmir Himalayas and

c) NS-trending Ganges-Brahmaputra graben and extensions of Monghyr-Saharsa Ridge in Sikkim and Darjeeling Himalayas.

Among the various types of preferred lithological associations, the volcano-sedimentary pile of rocks specifically of the lesser-Himalayan zone, host the most potential metallic mineral deposits. Some of their best examples include the Dalings of Eastern Himalayas (enveloping deposits of Rangpo and Dikchu in Sikkim, Genekha in Bhutan and Gorubathan in Darjeeling) and metavolcanics of Askot in the Kumaun Himalayas. In contrast to this, the Tethyan (eugeosynclinal) zone lying north of the Central Himalayan ridge is conspicuously poor in base-metal mineralization or, for that matter, in any metallic mineralization

(Krishna Murthy and Krishnaswamy 1976). The granitoids of Central Himalayan zone reveal many minor occurrences of base-metals and W, Sn, Mo, etc. but none of them has been found to be economically viable so far.

PROBLEMS OF EXPLORATION AND EXPLOITATION ACTIVITIES

In spite of intensive exploration for potential metallic mineralization in the Himalayas over the last 35-40 years, no major deposit of any metallic substance has been proved so far. The problems of exploration activities and of the mining of small-scale deposits have been studied with special reference to the polymetallic sulphide mineralisation at Rangpo, the only ore-producing mine of Himalayas at present. Briefly these are as follows:

Baffling controversies regarding numerous aspects of stratigraphy, structure, tectonics, metamorphism and metallogenesis pertaining to the mineralised units in particular (Ahmad et. al., 1976), Verma and Rao, 1984) generally circumscribe a meaningful exploration programme in many parts of the Himalayas. Stratigraphic and tectonic correlations and correlation-based estimates on grade and reserves often highly speculative with very low degree of reliability.

Apart from the conceptual problems, exploration activity in the difficult Himalayan tracts is generally beset with numerous other difficulties of techno-economic nature. The important ones among them are as follows:

(a) Difficulties in applying the usual direct guides of ore-search due to scarcity of observable outcrops and exposures of rocks/ mineralized zones except in few dangerous stream/river sections.

(b) Problems in distinguishing land-slided/transported blocks of earth from the in-situ occurrences.

(c) Generally irregular shape and size, very limited lateral and depth persistence and sporadic distribution of the ore-bodies.

(d) Difficulties in delineating the target-areas and estimating their potentialities due to lack of precise data regarding stratigraphic and/or structural correlations of the mineralized zones.

(e) Problems in the application of modern exploration techniques like geophysical and geochemical prospecting.The latter originate often due to heavy precipitation, steep ground-slopes, generally transported nature of solids, frequent land-slides,etc.

(f) Very limited scope for planned sequential-drilling and systematic exploration through definite grid-patterns of bore-holes.

As most of the areas around known occurrences are covered with thick and heterogeneous soil-and talus-cover, drilling through them always requires continuous casing of holes. For this reason, the drill-holes have to be reduced in size for runs beyond the overburden thereby decreasing the much-needed freedom and manouverability of drilling operations. Besides, the highly-disturbed nature of the drilling-ground often leads to heavy core-losses thereby reducing the reliability of the estimates of grade, reserves, etc.

(g) Problems related to frequent incidences of geotechnical hazards and hostile weather conditions apart from the lack of general infrastructural facilities, difficulties in transportation of men and materials, etc.

(h) Problems of beneficiating ores, particularly polymetallic sulfide ones (e.g. those of Rangpo, Dikchu and Gorubathan) specifically due to their complex textures, microstructures and mineral intergrowths.

At Rangpo, the constituent typomorphic minerals occur in such an intimate association and intergrowths that their desired liberation is not achieved even when the ore is ground to -300 size grade. Grinding of ores to this fine scale leads to heavy losses of metal values in the tailings and suspensions.

PROSPECTS OF SMALL-SCALE MINING

Inspite of the numerous problems and risks highlighted as above, the mineral resources of Himalayas deserve development on two important accounts : First, only these resources have the potentiality to be the harbingers of much-needed socio-economic progress in this vast stragically important, yet highly backward part of the country; secondly, the prospects of locating big metallic mineral resources possibly in depth still remain bright there. This is so in view of the well-established metallogenic potentialities of fold mountains mobilo-belts in many parts of the world like the American Cordilleras (hosting large prophyry-copper, lead-zinc, silver and tin depositss), Appalachians, Urals, Alpine mountain chain (with extensive Pb-Zn, Au-Ag, Hg, and multi-metal Ag-Co-Ni deposits) and South-East Asian Archepelago (with large tin, copper, au-Ag and Cu-Zn-Zb deposits). Although the plate-tectonic concept applied to Himalays (Guild, 1976) discourages prospects of finding big deposits, the question is:- should this be accepted as the final and uncontroversial global tectonic-metallogenic model applicable to Himalayas as a whole. If yes, why the picture is so different along the western extent of the same mountain chain particularly in Iran and Alpine Ranges of Europe? Huge metallic deposits like those of chromite in Baluchistan and copper ores in Iran are known to occur in the Western extensions of the Himalayan chain in Iran and Europe. These and many other questions remain still unanswered.

A planned strategy to explore and exploit the Himalayan mineral resources is called for. This requires first and foremost a basic philosophical change in the techno-economic considerations of resource development, the target area being a unique terrain with tremendous strategic and socio-economic significance. The doctrine of Intermediate technology professed by Dr.Shumacher (1978) of smaller working units, communal ownership and regionally-distributed work places utilising local labour and resources appears to be the best model applicable under the above-mentioned conditions. According to Barnea (1980) this approach suits the interests of the developing nations very well as the small mines can be brought into production quickly, thereby ensuring much-needed rapid returns of the capital. Moreover, they minimize social discontent and disturbances in the remote areas, channelizing the energy of the people for their own welfare and ultimately for the progress and development of the nation.

The doctrine referred to above, however, does not suit to the interests and work-ethos of the large public-sector mining organizations. The execution of small-scale mining operations in difficult and problematic conditions, as those prevailing in the Himalayas, may possibly of enterpreneurs possessing requisite drive and determination to achieve their goals with the same devotion as that of the mountaineers, in general. However, such enterpreneurs will have to be provided with considerable protection, encouragement and support by the Government. Of course, the government may have to enforce strict surveillance in the best national interest for strategic reasons. While such small units are allowed to concentrate on the actual job of mining, milling and marketing of the ores, the government can provide infrastructural facilities as part of the broader set-up of development of such strategic areas. Regional and detailed-scale exploratory operations that will form the back-bone of this strategy, will also have to be sponsored by the Government.

The long-term aspects of the strategy outlined above may possibly be best served through an integrated and multi-disciplinary approach in which geoscientists, mining engineers, environmentalists and even the planners of our defence capabilities interact evolving guidelines and plans to harness mineral resources of the mighty Himalayas to the best overall interest of the country.

No major discovery of any ore deposit having been reported in the last 10-15 years, it appears logical to conclude that we have possibly reached nearer to a point of super-saturation under the presently-pursued strategy of mineral resource development and it is surely the high time that we ponder over our failures analyse them more openly and change the strategies that have failed to produce results matching with investments. Goldschmidt, the Father of geo-chemistry advised against sticking too long to a method, technique or strategy that produced little or no result in a given framework of time. We possibly need heeding this advice sooner than later.

In formulating a new strategy or modifying the old one, we stand always to gain from the experience of successful nations. The model that appears most attractive to emulate at least in part, seems to be the one gifted by Goldschmidt himself, it is now followed throughout Scandinavia. The Geological survey organisations of these countries, having large untreadable mountainous tracts, have achieved a high degree of success in mapping large difficult areas, carrying out systematic research investigations and integrated mineral surveys and locating numerous workable mineral-deposits throughout the length and breadth of Scandinavian Caledonides through well-coordinated cooperation among the geoscientists from various research organisations, exploration and exploitation agencies and the universities. The strategy aims at avoiding unnecessary duplications and utilizing nation's precious resources-money, trained manpower and experts, to the best possible national advantage. Survey geologists working in close liaison withuniversities and research organisations, the selected ones directly involved in university research, training of manpower and even teaching, with their permanent offices in University teaching departments and the university Professors or scientists from the research organisations working with their dedicated band of students/researchers in field, laboratories and planning-cells with direct involvement in Survey's programmes, provide a strong basis for a well-knit cooperation and the best possible geological service that reflects in the high standards of their achievements in all spheres of mineral resource development. The key to success of this model is the well-coordinated engagement of nation's entire trained-manpower and expertise in a true spirit of cooperation and mutual faith, everyone contributing his might to achieve the clear lofty goal. There appears no reason why we cannot work similarly to achieve our clear goal too.

REFERENCES

Bernea, J. (1980): The Future of Small Deposits and Small - Scale Mining - in "TheFuture of Small-scale mining" - Published by UNITAR (1980), pp. 1-3.

Chatterjee K, et.al (1976): Tin-Tungsten mineralization in the Darjeeling Himalayas in Himalayan Geol. Seminar, New Delhi, Abstract Vol.P. 146-147.

Dar K.K. (1968): Metallogeny inthe Himalayas. XXII International GeologicalCongress, Mexico. Proc.Vol.7, pp.35-42.

Dasgupta, S.P. et.al.(1976): Sulfide mineralization in parts of the Himalaya, Himalayan Geology Seminar Section IV, New Delhi.

Gansser, A. (1964): The Alps and the Himalayas, 22nd Int. Geol. Congr. Proc.Vol.2, p.387.

Guild, (1971): Metallogeny and the New Global Tectonics Publ. McGraw Hills, New York.

Jhingran, A.G.(1974): Some studies of tectonics of Himalayas Intern. Sèminar on tectonics & Metallogeny of South-east Asia and Far East,, Calcutta, Abst. Vol. pp.3-5.

Krishnan, M.S. (1956): Geology of India and Burma Publ.: Singhbhum (P) Ltd., Madras.

Krishnamurthy, V.S. & Krishnaswamy, M. (1976): Volcanism and base-metal mineralization in parts of Himalaya in Himalayan Geology Seminar Proc.Sect. IV, New Delhi.

Mohammad, H., V.K.Gupta, O.P.Varma, S.H.Rasul and K.L.Rai (1976): Himalayas in relation to large-scale exploration programmes. Himalayan Geology Seminar Proc. Sect.IV. Abst.Vol.pp.159-160.

Mukherjee, N.K., Lahiri, D. & Bhaumic (1976): Problems and Strategy in development of basemetal prospects in the Himalayas. Himalayan Geology seminar, New Delhi (1976) Section IV, paper 25.

Schumacher, F.F. (1974): 'Small is beautiful: A study of economics as if people mattered'. ABC Publications USA.

Varma, O.P. (1986): Some aspects of ultramafic and ultrabasic rocks and related chromite metallogenesis with examples from Eastern Region of India Presidential address, 734d Ind.Sc. Cong.Geol.& Geog. Section, New Delhi, 72 p.

Varma, O.P. & Rao, A.M.S.(1984): Some aspects of structural setting, metamorphism and tectogenesis of some base-metal deposits and their associated host-rocks in the Eastern Himalayas. Seminar on 'Himalayan crystallines, metamorphics and structuresS'. Abst. Vol.pp.38-41.

Verma, R.K. et al (1976): Seismotectonics of the Himalaya and implications of continental plate convergence. Himalayan Geol. Seminar, Abst.Vol.pp. 116-117.

Wadia, D.N. (1938): The structure of the Himalayas and of the North Indian Foreland. Indian Science Congress Proc. Vol.XXV, Pt.2, Calcutta, 1988.

Wadia, D.N. (1957): Geology of India, McGraw - Hills Publishing Company Ltd., New Delhi.

TABLE 1: SPATI AL DISTRIBUTION OF METALLIC MINERAL DEPOSITS IN THE HIMALAYAS

TECTONIC ZONE	WESTERN SECTOR		CENTRAL SECTOR		EASTERN SECTOR		
	JAMMU & KASHMIR	HIMACHAL PRADESH	UTTAR PRADESH	NEPAL	BARJEELING SIKKIM	BHUTAN	ARUNACHAL PRADESH
Trans Himalaya							
Tethys Himalaya	Native copper-Zanskar (Ophiolites) (Ladakh)						
Greater Himalaya			Base-metal -Galapkot	Base-metals Budhakhola Nangri Khokling			
Lesser Himalaya	Base-metals -Buniyar	Antimony-Barashgri (Precambrian) Copper-Naraul- -Dhanaula Uranium-Kulu (±Cu)	Cu-Rainagar Pb-Zn-Cu-Askot -Chhanapani - Shiskhani Pb-Pokhri	Base-metals (Matha-gwar)	W-Monsong(WB) Zn-Pb-Cu -Rangpo -(Sikkim) Cu-Zn -Dikchu (Sikkim) Pb-Zn-Cu -Gorubathan (W Bengal)	Pb-Zn- (±Cu) Ganekha and Chakula	Co-Ni-Cu Potin Yazakli
Sub-Himalaya (outer Himalaya)	Bauxite-Riasi	Copper-Uchich (Red beds)	Uranium in phosphate rock - Mussorie syncline			Placer Gold - Kalapani	Placer Gold - Subansiri

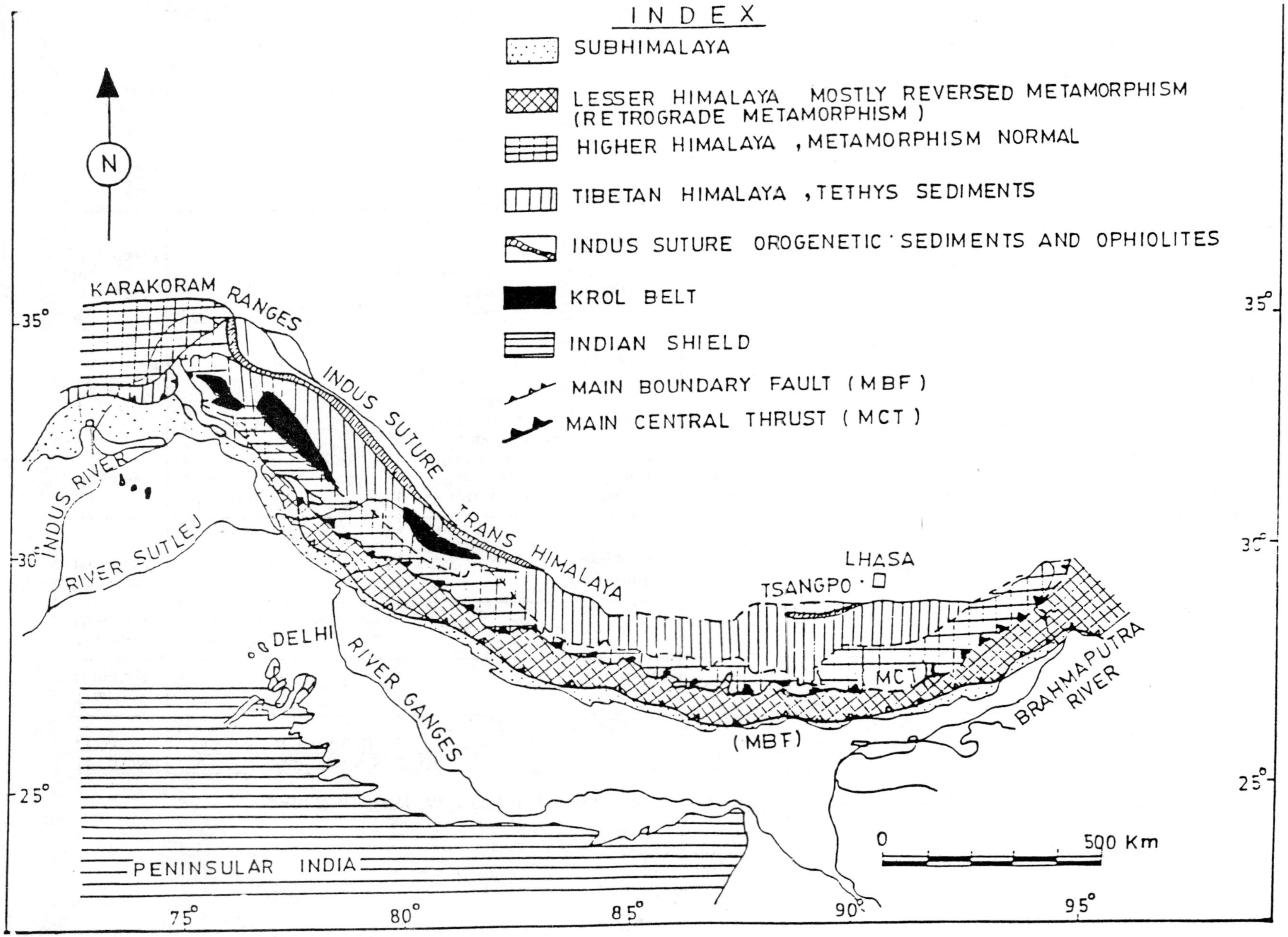

FIG. 1 LITHO-TECTONIC ZONES OF THE HIMALAYA (Modified after Gansser, 1974)

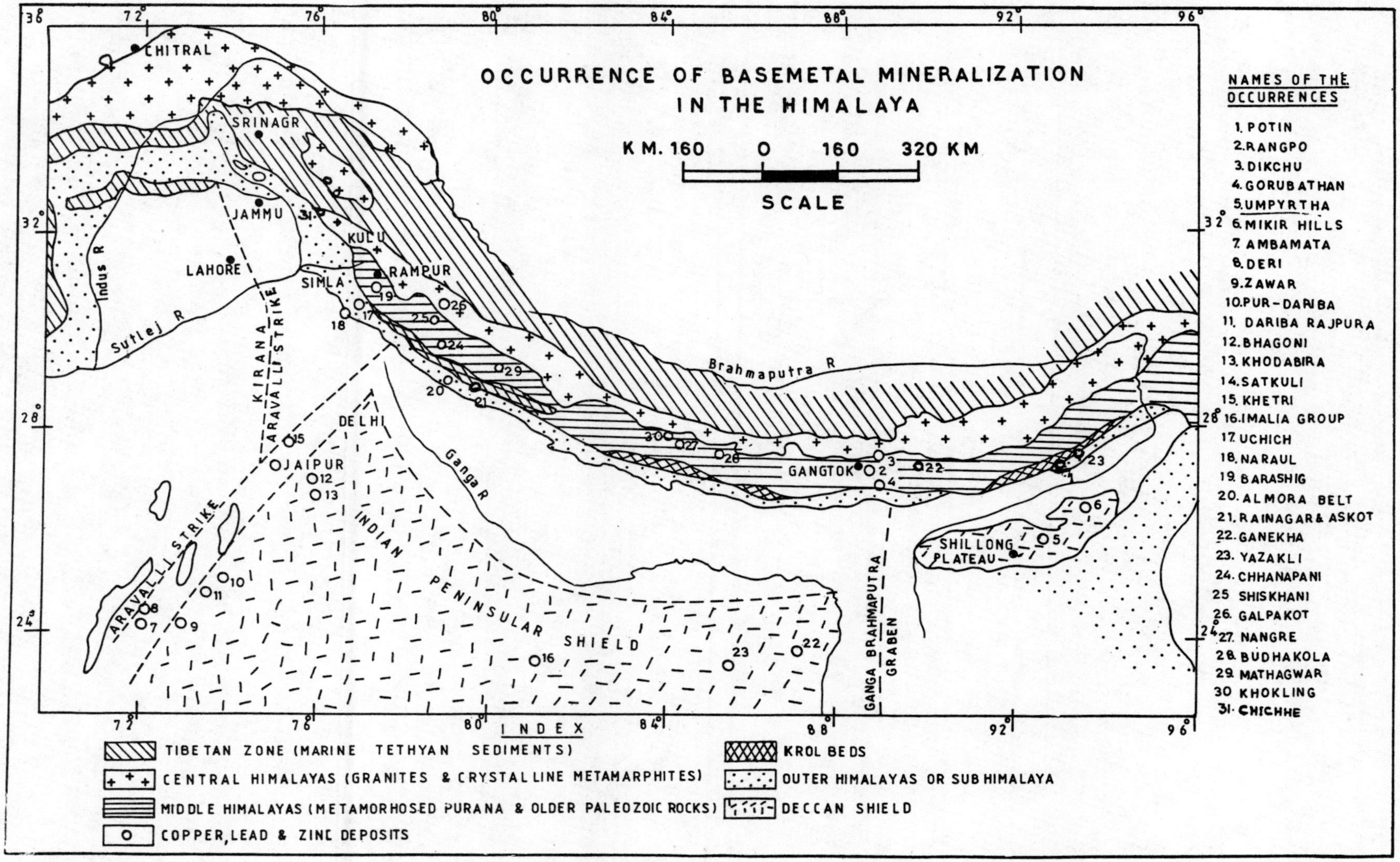
OCCURRENCE OF BASEMETAL MINERALIZATION IN THE HIMALAYA
KM. 160 0 160 320 KM
SCALE
NAMES OF THE OCCURRENCES
1. POTIN
2. RANGPO
3. DIKCHU
4. GORUBATHAN
5. UMPYRTHA
6. MIKIR HILLS
7. AMBAMATA
8. DERI
9. ZAWAR
10. PUR-DARIBA
11. DARIBA RAJPURA
12. BHAGONI
13. KHODABIRA
14. SATKULI
15. KHETRI
16. IMALIA GROUP
17. UCHICH
18. NARAUL
19. BARASHIG
20. ALMORA BELT
21. RAINAGAR & ASKOT
22. GANEKHA
23. YAZAKLI
24. CHHANAPANI
25 SHISKHANI
26. GALPAKOT
27. NANGRE
28. BUDHAKOLA
29. MATHAGWAR
30 KHOKLING
31. CHICHHE
CHITRAL
SRINAGR
JAMMU
LAHORE
KULU
RAMPUR
SIMLA
DELHI
JAIPUR
GANGTOK
SHILLONG PLATEAU
Indus R
Sutlej R
Ganga R
Brahmaputra R
KIRANA ARAVALLI STRIKE
ARAVALLI STRIKE
INDIAN PENINSULAR SHIELD
GANGA BRAHMAPUTRA GRABEN
INDEX
TIBETAN ZONE (MARINE TETHYAN SEDIMENTS)
CENTRAL HIMALAYAS (GRANITES & CRYSTALLINE METAMARPHITES)
MIDDLE HIMALAYAS (METAMORHOSED PURANA & OLDER PALEOZOIC ROCKS)
COPPER, LEAD & ZINC DEPOSITS
KROL BEDS
OUTER HIMALAYAS OR SUB HIMALAYA
DECCAN SHIELD

FIG. 2

MANAGEMENT PROBLEMS AND REMEDIAL MEASURES FOR TECHNICAL CHANGES IN THE MINERAL INDUSTRIES OF THE LDCS (WITH PARTICULAR REFERENCE TO INDIA).

Prof (Dr) D K Sinha
Director
Indian School of Mines, Dhanbad

I. INTRODUCTION

Industrial Development in India in the last few decades of this century have been identified with a remarkable growth of mineral production to cope with the demands of national economy as well as the world export market. Traditionally, Mining Industry of these countries particularly in India was tuned to semi-mechanised and manual systems of mining organised through smaller size organisation and now it is faced with a challenging task of adopting the higher technology which has been necessary for intensifying large scale production through large and complex organisation. Technological and organisational changes have therefore been inevitable in the present day context which are accompanied with various management problems which have been tended to decelerate the desired growth rate resulting in lower overall efficiency.

A review of the current status of the technological system practised in Indian mines would show in most of the situation the technological practises are of lower order and in some cases where improved higher technology have been imported and introduced, the results are not upto the expectations. Unfortunately, overall experience gathered in this area during the last one ana a half decades shows that introducing higher mining technology in Indian conditions has not been very encouraging because of low performance and productivity compared to that of the advanced countries of the world. It is a fact that for an industry any change either in technology or in the organisations is bound to bring certain associated problems which need to be correctly identified and remedial strategies must be planned to minimise the threats of slow progress and negative overall productivity. The organisational systems consist basically a few sub-systems such as technical, structural, psycho-social, goal and value sub-systems which are interrelated to one another and any change in any one or more of these subsystems would demand a change in the management system which is the key sub-system integrating and coordinating the functions of these sub-systems. The management systems need to be dynamic and re-designed with the changing one or more of the above mentioned sub-systems and the relating the socio-economic environment controlling the system. Inevitable technological changes in the mineral sector have, therefore, brought about organisational changes and the associated management problems which are highlighted in the paper are based on the author's experience in the above mentioned background.

II. Major Management Problems arising out of the Technological Changes.

1) High investment accompanied with low return. A developing country is always capital scrace where the problem of consistent low ROI is devastating to the economy.

2) Low out-put per machine shift of operations comparable to similar situation in other countries.

3) High resource consumption in terms of man-power, materials and money resulting in lower utilisation factors.

4) Low overall productivity arising out of the above mentioned points, particularly a terribly low out-put/ capital ratio which a developing economy can hardly sustain for a longer time.

5) Lack of effort for effective absorption of technology where experience gained in some typical mining conditions is not utilised profitably at some other conditions. This is due to lack of proper data-base and effective use of information technology.

6) Conventional Management Style inherited from the traditional social and political systems of the past which does not fit into the dynamics of today's social change.

7) A typical resistance to change of the management style by both bureaucrats and technocrats.

8) Lack of motivation for higher performance on the part of both the workers and managers.

III. Cause-effect Analysis of the Problems and Approach for Remedial measures.

1) Selection of technology - right type of technology should be selected for classified typical geomining conditions of deposits which would ensure optimum productivity in terms of out-put per man day, and return on investment. Based on analytical studies carried by the author in the Indian mining situation, methods for selection of "optimal technology" taking into account the conflicting criteria of output, investment and man-power, has been worked out and applied in some mineral basins. Thus optimal technology can be chosen for given mining situation using quantitative techniques of analysis and computer methods. The computer models have been developed on the basis of an Operation' Research techniques using Integer programming and Dyamic Programming methods for selection of technology on a longer term basis.

2) Once a optimum technology is selected, a suitable operational management system is developed based on computation of standard capacities for different types of machines in different situations and standard manpower computed so that the capacity is maximised in terms of available man hours and machine hours. Industrial Engineering techniques including computerised techniques and maintenance management would be of much use in this area. Fairly extensive research work is in progress by the author for the operational management system optimisation.

3) For a high technology mine, the organisational system should also be restructured giving due regard to the specialisation for different management staff position, specialised training needs are to be established in different functional areas of management which would ensure the services of professional managers in the industry. Further organisational development including team building techniques would take into account the dynamics of the organisational change in the context of the renewed goal and value system of the new organisations.

4) The social system approach to management need to be stressed which is the talk of the day. The Western and Eastern styles of management have been developed based on the social tradition and culture. We have seen the success of the Japanees style of management in that country, the Western style of management have been successful in those countries whereas countries like India having widely diversified tradition and culture an indigenous style of management need to be evolved to fit the social tradition and culture.

Whatever technological innovations have been done and introduced, it is the people who are the key to progress. Ethics, values and the aspiration of the people in developing countries have been undergoing a rapid change which the Govt is expected to fulfil. The identification of this change and designing the organisational system accordingly is the main task of today's management. In developing countries like India people may be poor but have proved their worth time and again and their capability in terms of creativity, craftsmanship, ability and endeavour. It requiresa suitable industrial environment to motivate and orient their endeavours towards achievement of better goals. Therefore, as managers we need to examine our own attitude towards our human assets which would help in mobilisation towards higher productivity and excellence.

IV. Conclusion

Technological changes and modernisation of operations are two unavoidable phenomena for growth of mineral industry in the developing countries. In order to achieve the desired growth rate effectively, the associated management problems of change are briefly discussed in the paper. This can be classified into two - the change in the system and the change in the environment. Change in the system is the outcome of deliberate decisions of the management to increase production to satisfy the demands of the consuming sectors; whereas the change in the environment is a natural outcome of the Govt's political system which is accompanied by the values and aspirations of the people. Therefore, the management system in the mining sector needs to be designed in a manner that it can integrate the environmental supra-system of the nation and the technological and corresponding organisational sub-system of the company. This will be only possible through a formal management education developed in the lightof the socio-economic situation of the concerned nation, its past tradition, culture, ethos and value system. A new management style which fits best into the demanding nature of the changes in the technical and human system will have to be established and practised.

MINERAL DEVELOPMENT IN INDIA
-POSSIBILITIES, PROBLEMS AND STRATEGIES

S.K.Sarkar
Central Mining Research Station
Dhanbad

INTRODUCTION:

Contribution of the mineral sector to the material progress of a country is immeasurable. Throughout the history of development of a country discovery of a new mineral or a new method of extraction, and processing of mineral, has a major impact on industrial growth. Substantial benefits can accrue to a country from a well developed and well administered mineral sector. However, the development of mineral sector in any country whether developed or developing forms only a part of total national development and must therefore fit with the framework of total economic development plan of the country concerned.

Presently, India produces more than 50 minerals (excluding atomic minerals) from more than four thousand mines scattered all over the country. There are a large number of small mines for various minerals. The size of mines in fuel sector and non ferrous and precious metal mines is however relatively larger. By and large small mines are mostly either proprietory or partnership venture. Fuel mineral, precious metals, Diamond and a number of other minerals like lead, copper pyrites, phosphorites are produced exclusively by Public sector whereas in the remaining minerals public sector shere vary between almost nil to about 90 percent.

The approximate sizes of mine in terms of annual production in different sectors are as follows:

Coal (including lignite)	- 283,000 t
Metal mining	- 68,000 t
Non metal mining	- 15,000 t

OBJECTIVES

The dominant objectives of mineral development plan of a country would depend on the country's needs. The dominant objectives of the mineral development of India may be laid down as follows:

1. Assuring adequate supply of raw material for the country's industry.
2. Ensuring optimal use of indigenous mineral resources;
3. Creation of employment directly or indirectly.

4. Earning or saving of foreign exchange.

5. Stimulation of regional development.

The country is self sufficient in a large number of minerals including Iron Ore, Bauxite, Chromite, Manganese, Limestone, Magnesite and ferro alloys. In all these minerals the production is more than the domestic requirement. The country's requirement of copper, lead, zinc, Borax, Sulphar, Asbestos and Rock phosphate are however not met from indegenous sources. The deficiency is more than eighty percent for Borox, Sulphar and Asbestos whereas for copper lead and zinc it varies between approximately 45 and 75 percent.

The creation of employment opportunity should be one of the dominant objective in country's mineral plan. In the seventh five year plan, creation of employment opportunity has been mentioned as a vital objective and it should be so considering the vast unemployed manpower in India. The mineral industry has got a potential for creation of employment opportunities in remote areas. The Mining and quarrying division occupied seventh position among the mine National Industries Division at 1982. The total employment was 961,000 persons which meant 4.9 percent of total employment as at 1982.

As an earner of foreign exchange for the country the mineral industry has got an important role to play. In 1972 the total foreign exchange earning from mineral sector was 2368 million rupees which has increased to 23968 million rupees by 1982 accounting for about 29 percent of total export of the country. The crude and semi-pressed/processed minerals contributed about 96 percent of total mineral export whereas mineral based manufactured products accounted for the rest. The balance of trade position however changed during the period 1972 to 1982. In 1972 the export was 106 percent of import bill whereas in 1982 minerals export accounted for about 47 percent of total import bill.

TECHNOLOGY:

Present Status

During the past decade, there has been a world wide trend towards massive mining operation to take advantage of the economics associated with large scale production and use of large mining equipment. For example, in coal mining, the longwall faces are now being planned for a production capacity of 8 thousand tonnes per day under the advanced mining technology programme of USA. In open cast, a production capacity of 25 to 30 Million Tonne per year and an over burden ratio of 1 in 25 are not uncommon. In case of hard rock mining, the typical 40T truck has given way to 200 T truck. The Shovel size has also increased tremendously. The major objectives of resorting to large scale mining and processing world over is to reduce cost and to find a way out from depletion of number of labourers willing to work underground.

In India, the technology scene is not uniform as it happens, perhaps, in other countries too. The public sector undertakings like Coal India or Indian Copper have tried to increase the scale of their mining operation. Coal India is planning for open cast projects with an annual production capacity of 3-5 million tonne, and with an overburden ratio of 1 in 6. For the underground mines, there has been consistent efforts to introduce advanced technology longwall mining on a large scale. The number of mechanised longwall faces are proposed to be increased from 8 at 1986 to more than 100 by the turn of the century. But reviewing experiences of modernisation of mining operation vis-a-vis cost reduction, it may be said that whereas large open cast projects have definitely demonstrated their economic viability but the same cannot be told about the underground mining operations. Uptill now, underground modernisation programme in coal mines have not been able to produce coal at a cost cheaper than that attained in similar condition with conventional nonmechanised and semi mechanised technology.

The idea, however, is not to belittle the role of advanced technology or modernisation in the Indian mining scene. There are difficult situations at a great depth, and with adverse geomining conditions where modernisation is a must. But for a large number of small coal and non-coal mines scattered in far flung area of the country, modern and massive technology may not be the most appropriate technology. Moreover, the capacity of such units to absorb advanced technology would be very low. A coal mine situated in a remote area where necessary infrastructural facilities like rail and road are not adequate, would not be able to sustain mechanised longwall mining. Actually, with prevailing labour cost and average geo-mining conditions, there is an optimal size at which the cost may be minimum and that size may not be very big or even big as per international norms. Summarising, it must be said that in the prevailing mineral industry scenario of India, a mix of technology is a must. For open cast projects and underground projects at great depth we may go for most modern technology, but in the thousands of small mines scattered all over the country our endeavour should be to reduce human drudgery and manual labour to the best extent possible and to go for an optimal programme of modernisation which will give the mineral at an acceptable cost.

Future Technology

Nodule Mining

One of the frontier area of mining technology which has got a potential for future is the exploration and exploitation of Ocean resources. Of the 14 or 15 minerals most pleantiful in Sea water, only 3 have been recovered in large amount up to late 1970s. These are Salt, Magnesium and Bromine. Commercial exploitation of mineral resources within bed rock (Sub surface deposit) is uptill now limited to Oil and Gas and a small group of minerals including coal, sulphar and iron ore.

But from the future economic stand point of view, the nodule mining from the Ocean resources has attracted world wide attention. Nodules are metalic objects which vary in size from a pea to a base ball and are found scattered over a large area of Ocean floor at a depth from 900 to 6000 m. The metalic composition vary widely. But a typical deposit of commercial mineral contains 25 to 30% magnesium, 1.5% nickel, 0.5 to 1% copper. Although the existence of nodules has been known for about 100 years, the widespread interest for their exploitation and exploration is of recent origin.

There have been 35 major companies upto 1980 engaged in exploration, development and equipment construction for Nodule Mining in the USA, Japan and West Europe.

Though the economic viability of Nodule mining is a far cry at the present stage, India, alongwith a number of other developing countries have taken lot of interest in the programme of nodule mining and the associated conferences convened by the United Nations. The National Institute of Oceanography under the Council of Scientific & Industrial Research alongwith a few other Government Agencies like I, IBM, GSI, HCL, HZL, and NGRI have done some experimental work on Surveying of polymatallic models in Central Indian Ocean. The Surveys comprise echo sounding, sea bed photography and sea bed sampling by grabs. Experimental work have been undertaken in the laboratory to develop processes for winning metals like nickle copper cobalt and manganese from the ocean nodules.

Recovery & Recycling

Another technology which has got a potential for future is the recovery and recycling of waste from previous mining operation. Mineral waste especially sub-marginal tailing dumped near mining site is likely to increase with every passing year and possible recycling of this at a future date may be an attractive proposition.

The IBM has carried out a number of studies for recovery of ore from tailings. The CMRS has conducted experimental studies in the laboratory and the field on bacterial leaching of low grade ores. The studies are however in experimental stage but the low grade failing from mine may be a source of supply of mineral in future.

CAPITAL REQUIREMENT

During the first half of the century when mining was being carried out on small scale, the industries were virtuallyself-financing, obtaining a very small percentage of its capital requirement from external sources. With increase in massive and modernised operations, the capital requirement of the mineral sector has enormously increased. But the general psychology still prevailing is that a mining is a high risk venture. Naturally when such a high risk industry compete for fund with risk industry compete for fund with lower risk industries, it has certain disadvantages. A large part of mineral activities in India is by the

Central & State Government Undertakings. Their fund requirement is looked after by the Governments. Some mineral companies though not very frequently have issued bonds in the market to secure loans too.

The Government, however, has secured some soft loans for different mining projects from bilateral agreements with different countries. The World Bank and United Nation's development programme (UNDP) have also occasionally helped. Upto 1976 the World Bank has financed nearly 750 million dollars to mineral projects which accounted for about 23% of total project cost. The loan was provided either in the name of the World Bank or through its subsidiaries International Development Association and International Finance Corporation. Roughly, 45% of the loan was for Africa, 37% for Latin America and 9% for India. The loan provided to India during the period included 35 million dollars to private sectors coal production (1961) and 19.5 million dollars to Chasnalla coal mining projects (1961). The United Nation's Agency, Centre for Natural Resources, Energy and Transport, on behalf of the United Nation's Development Programme had contributed some finance for mineral sector development in developing countries. The finance has mostly gone to mineral exploration programme using modern methods establishment of mining institutes and other technical assistance activities. The total UNDP assistance upto 1976 was Rs.132 million dollars of which approximately 100 million was for exploration and exploitation whereas the remaining was for Institution building and strengthening. During the period mentioned the total UNDP contribution to India was 2905 thousand dollars for institute building. The above exposition underlines the need for investment in mining sector from internal sources. The commercial banks should keep a proportion of their loan for minerals industries only. The proportion should be fixed on the basis of mineral industries shere in total GNP of the country. The loan may be channeled through separate mining windows of some of the selected commercial banks having base in mining areas. This loan would specially be useful to private mine owners operating on Small Scale.

ENVIRONMENT

The environmental problems caused by industries have become a significant issue in India recently. The citizens in surrounding areas of many mining projects (as in other projects too) have voiced their grivances to authorities regarding the problems such projects created or might create. It is now the declared policy of the Government to take opinion of the Ministry of Environment regarding the environmental problems a project may create before finally clearing a proposal.

The mining operations because of its very nature create some special environmental problems besides the problem of dust and smoke common to all the industries. These problems include deforestation and change in land profile due to open casting, subsidence due to underground mining, .damage due to blasting vibrations and lowering of water table due to increasing depth of underground mining operations.

In future scheme of things, land reclamation after opencasting should form a past of the project operations, wherever possible. This is easier said than done. The large mining companies even if they do it would have to pass on the cost to the consumer which in turn, may have financial implication on the project. But the public awareness on the subject has been raised to such extent that in most of the cases in future it would be better to plan reclamation as a part of the project. However, if Land reclamation is enforced through some legal measures many of the small scale mining operations would have to wind up. In the future a compromise (between economic requirement of a project and environment of the surrounding area) is necessary and to achieve this compromise local community should be involved in the problem.

Increasing number of complaints regarding damage caused to building and other structures due to blasting vibrations are now being received. The limit of vibration upto 50 mm/sec (Particle velocity) at present being allowed is perhaps based on insufficient observations. Cases are on record where authorities in U.K. have stipulated that in built up areas the particle velocity of vibration should not exceed 6 mm/sec. Further study on the subject under Indian conditions is a must.

Water scarcity caused by lowering of water horizon due toextensive mining is another environmental problem which has not attracted the attention it deserves.

To an interested observer in Jharia and Raniganj coalfield it would be apparent that number of wells and water tanks getting completely dried during summer months have increased. The duration over which these sources remain dry have also increased. In future scheme of projects specially in areas with large concentration of underground mines, provision of water supply should be planned not only for the mining colony but to adjoining non mining population not covered by any scheme of municipal water distribution.

TRAINING

In recent years, great effort has been put on the training of personnel for mining industry. Compared to other developing countries, India does not suffer from dearth of adequate qualified manpower. But that does not belittle the scope of training. At present a large number of courses are available for programme upto masters and doctorate level in India's Universities and Institutes. Under different bilateral programmes and also under a number of United Nation's programmes, quite a large number of fellow have been sent to foreign countries. While all these helped us todevelop the quality of our manpower, a few remarks given in the following paragraph seem to be pertinent:-

(i) The maximum stress has been put on the training of professionals but, perhaps mineral exploration and exploitation programme may benefit a lot if sub professionals also are sent to other countries for on the job training. It is alright that senior personnels like Managers, Agents and above should be exposed to training and advanced courses in industrialised countries, but for ten professionals sent, we should send atleast 5 sub professionals like Electrical Foreman, Senior Technicians and Overmen for 'on the job' training.

(ii) It is frequently observed that an Officer who has been exposed to a training in foreign country is not put in a position where he may utilise the training he achieved. As for example, an engineer who has been trained in longwall technology, may find himself transferred to an area where there is no programme of longwall in the foreseeable future. The entire training programme should be so designed that such type off pit falls do not occur.

(iii) There should be exchange of persons between educational universitues, research institutes and industries. That will help everybody to appreciate each others' problems better.

SMALL SCALE MINING

Small scale mining deserves separate mention in any discussion of Indian mining scene. The production level at which a mine should be termed as small may vary from one country to other. But such mining is generally characterised by labour intensive and inefficient work method, inefficient exploitation of deposit, sub-standard work condition and undependable output. Neverthless, small scale mining makes important contribution to the country's economy as would be seen that in 1982 total value of minor minerals which are mostly by small scale mining is 30% approximately of total value of mineral produced excluding fuel minerals. In this connection, it should be mentioned, however that not only minor minerals but a large percentage of non-metal mines andeven quite a modern percentage of coal mines may be characterised as small scale operation. The major problems of small scale mines in our country are the following:

(i) A small lease hold prohibit Scientific exploitation of mineral resources.

(ii) Inefficient mining methods mean that only higher grade ores are mined which causes substantial losses of national mineral resources.

(iii) Operations are generally so small that provisions of adequate infrastructural facilities become difficult.

(iv) Work condition and health standards are very poor.

(v) Mine operators have very limited data about the reserves and its quality.

(vi) The financing of such operation for organising it in proper way is difficult. Because of the uncertainity and risk involved in small scale mining finance may not be available on favourable terms.

The Government has taken some steps to alleviate some of the problems of the small scale mining sector but much remains to be done. The facilities for exploration, planning, and analysis should be further extended. There is scope for a separate agency or a completely independent cell in an existing agency to undertake the planning and exploration function on behalf of the small scale mines. The special credit facilities for the sector may be created by establishing a special mining bank or opening a mining window in the commercial banks operating in areas where small mines are concentrated. Many of the small mines are too small to justify the purchase of equipment for exploration, drilling and development. The Government may consider the creation of an equipment pool for use on loan basis by small scale mines. A separate agency or a separate department under some of the existing organisation in the field may cater for the job.

Technical assistance for training of miners in mining methods and atleast some of the available engineers in planning and managing functions may also be arranged by the Government.

Recently, there has been complaints that small scale mining operations being conducted some time even do not have record of persons employed and safety standards are never adhered to. In this connection, it may be worth considering whether in few cases and on experimental basis small scale mining of deposits in far flung areas may be handed over toe the local district council which may operate the mines in commune mining pattern of China.

ON MAXIMIZING THIRD WORLD BENEFITS FROM MINERAL RESOURCES

by Ann Seidman
Brandeis University, USA.

I am grateful for the invitation to participate in this International Symposium on "Strategies for Exploitation of Mineral Resources in Developing Countries." Scools like yours will undoubtedly play a key role in devising new strategies for exploiting mineral resources as a foundation for building self-reliant, integrated development capable of providing productive employment opportunities and raising the living standards of third world peoples. I welcome this opportunity to share information and ideas.

THE UNDERLYING PROBLEM:

At this symposium, I need not dwell on the nature of the historically-shaped problems confronting minerally-based economies today. You are all as aware as I as to the way, for centuries, foreign firms employed local labour at low wages to mine third world countries rich minerals and ship them for processing to their factories back home (e.g.,A.Seidman,1975). The parent companies based in western industrialized centers not only owned the mine, they also manufactured mining equipment and machinery, and handled the marketing and processing of the mines' for output for the final consumers.

The peoples of host countries kbenefitted little from the multiplier effects promised by mainstream economists. Enclave-like mine development generated few backwards or forwards linkages to other sectors of the national economy. Only a small fraction of the labour force obtained even low paid jobs on the mines -- and in recent years increasingly capital-intensive mechanization has reduced that fraction. Few workers learned the skills needed to manage the mines, far less increase productivity in other sectors of the national economy. And the companies remitted a major share of the profits to their home countries -- in small countries like Zambia and Namibia as much as 25 to 40 percent of the Gross Domestic Product -- leaving behind little capital to start other productive activities. The host countries spent the little foreign exchange they earned for their mineral exports to import high priced manufactures for those few who could afford to buy them. When the world prices for crude minerals fell, the foreign mine companies simply laid off workers and cut back production-- and crises spread through the national economies.

After World War II, nationalist movements held out the hope that independence would assure a greater share of the benefits from their countries minerals. But, as we all know too well, their bright dreams dashed on the rocks of the harsh reality. Even when newly independent national governments acquired shares in their countries mines, seeking to take over the management and capture the investable surpluses generated, the majority of their peoples benefitted little.

By the late 1970s and early '80s, the international crisis, engulfing most third world mining economies, exposed their relative lack of success in contributing to greater national welfare. In the face of relatively inelastic demand, their competition to expand the mines output, together with the introduction of substitute products, pushed down world mineral prices. They borrowed heavily abroad at high rates of interest to cover the on-going costs of their expanding bureaucracies, social services and other development programs. As Table I shows, debt service payments began to consume an increasing share of their foreign exchange earnings. The International Monetary Fund (IMF), in return for its assistance, imposed an austerity package that typically contributed to growing unemployment, reduced social. services, and falling real incomes (A. Seidman, 1986a; Makgetla, forthcoming). At the same time, the IMF pressured their governments to halt even the limited 'interference' through which they had sought to augment their mining sectors' contribution to national welfare.

Table I: EXTERNAL DEBT OF SELECTED MINING ECONOMIES

in millions of U.S. dollars and as percent of GDP

1970 and 1983

Country	External public debt (millions of dollars)		As percent of GDP	
	1970	1983	1970	1983
Bolivia	479	2,969	33.8	77.7
Chile	2,066	6,827	25.8	39.2
Guinea	314	1,216	47.4	69.2
Jamaica	160	1,950	11.8	65.2
Liberia	158	699	49.6	72.1
Papua New Guinea	36	911	5.8	40.4
Peru	856	7,932	12.6	48.1
Zaire	311	4,022	17.6	91.5
Zambia	623	2,638	37.0	83.9

Source: World Bank Annual Development Report,1986
(New York: Oxford University Press, 1986)

The question is, why? How can developing countries devise effective measures to ensure their mineral wealth contributes more to spreading increasingly productive employment opportunities and rising living standards for their peoples?

AN EXPLANATORY HYPOTHESIS:

I would like to suggest an hypothesis to explain why so often in the past, despite bold declarations at independence, developing countries have failed to attain their proposed goals. To the extent that it is valid, it suggests anagenda for the kind of research needed to design more successful policy measures.

My hypothesis comprises two basic propositions. First, despite their declared determination to exert greater national control over their mining sectors, most post-independence governments largely left intact colonial-shaped institutional structures that delegated primary decision-making power to foreign mining firms. In the context of the post World War II technological revolution that multiplied the capital cost of mining machinery and equipment, transnational corporations asserted even more concentrated control over international sources of mining capital, technology, managerial personnel and markets (Seidman and Makgetla, 1980). Despite nationalist governments marginal changes in national institutions, these still functioned primarily toreplicate inherited patterns of mining investment. They left the host economies dependent on crude low-cost mineral exports directly or indirectly controlled by a handful of transnational mining conglomerates.

If valid, this first proposition focuses attention on the need to investigate alternative ways of exerting more effective national control over mining sector institutions.

The second proposition contends that, even where they succeeded in exerting a degree of national control over mining sector institutions, most governments failed adequately to plan the integration of the mining sector into national development. Too often, the mines continued to expand as enclaves, aggravating the dualistic external dependence of the national economy while underdevelopment persistently impoverished the mass of their populations.

If valid, this second proposition argues for governments to formulate and implement plans to integrate the mining sector into all aspects of their longterm industrial strategies. Only then can they ensure the nation's mineral wealth will contribute more effectively to the people's welfare.

To substantiate this hypothesis, I will summarize the evidence as to why, first, third world governments did not more successfullyexert national control over the mining sectors; and, second, they failed to incorporate the mines adequately into their overall development strategies.

1. Government efforts to exert national control:

For this audience, I will not waste time debating time-worn mythologies about whether a government should intervene in the mining sector(1). We all know no markets can exist "free" of the exercise of government power (Myrdal, 1972) Governments always intervene, either directly or by allocating decisions to private actors (R.B.Seidman,1978). They shape the framework and define the conditions within which private "market forces" function. The fundamental issue is, not whether the government should intervene, but how? To what institutions, and under what circumstances, should government -- as politically-organized society's only instrument for social change (Chambliss and Seidman,1982) -- allocate decisions relating to the mines in order to achieve national development goals?

Newly independent governments devised various measures to enhance national control over their mining sectors (A.Seidman, 1975). Many aimed primarily to increase the government's share of the investable surplus generated by the mines to finance their development plans. Some sought improved worker training,

---

(1) I will not, here, discuss the critical issue of which interests, in a given society, the state itself represents; self-evidently, however, if it does not represent those groups that stand to gain, it will not pursue policies or reshape institutions in ways likely to achieve more balanced, integrated development.

processing of the crude minerals before export, and the creation of backwards and forwards linkages with the rest of the economy.

To explain why these efforts so frequently failed to achieve governments declared goals, however,requires analysis of the behaviour of those managing the mine companies in the context of the decision-making structures that shaped policies relating to key issues like: Which deposits to mine? How many workers to employ and what to pay them? Whether to use labour or capital-intensive technologies? Whether to encourage local manufacture of machinery, equipment and component parts or to import them, and how much to pay for them (including whether to pay high prices to overseas affiliates to transfer profits, undetected, out of the host country? How much to produce in any given period? To what companies and at what price to sell the output?

How the responsible personnel behave in formulating decisions in response to a new government directive depends on many factors. These include whether they receive and fully comprehend the directive; whether, given objective constraints, they have the capacity to conform; whether they consider it compatible with

---

(2). Studies concerning what determines the behaviour of relevant individuals in response to new laws have systematized these factors under the mneumonic, ROCCIPI: Given a Rule, do they have the Opportunity to obey? has the rule been adequately Communicated to them? do they have the Capacity and do they their personal interests and their world outlook[2].

As in all large private and public institutions, the input, conversion and feedback processes that characterise the mining firms decision-making structures influence the behaviour of the key policy makers. The input processes include the channels that determine the kinds of issues and information the decision makers consider. Historically, in third world mining sectors, transnational corporate engineers, marketing experts and managers decided through which channels to obtain these inputs.

The conversion processes consist of the institution's internal standard operating rules and procedures (c.f. Allison, 1970) which determine which decision-makers -- e.g., boards of directors, managers or technical experts-- make particular decisions and by what criteria. In the past, in the typical third world mining economy, company managers, rather than board members, made the crucial day-to-day decisions in accord with overarching operative criteria relating to maximizing the parent firm's global profits.

The feedback proceses govern whether or not and through which channels those affected by a particular policy may express their reaction to it.

---

perceive obedience to be in their Interest? Does the decision-making Process foster their obedience? Do their Ideologies, their values and attitudes, support conformance? (See R.B.Seidman, 1978)

Traditionally, in the typical mining sector, these permitted responses primarily from top managerial personnel, financial institutions, and parent corporations, as well as price signals from world markets. Hierarchical managerial structures typically excluded workers' feedback, leaving them little choice but torely on demonstrations and strikes to protest particular policies they opposed. Small national business interests and even host governments had limited access to corporate policy makers.

Figure 1: FACTORS INFLUENCING DECISION-MAKING WITHIN INSTITUTION:

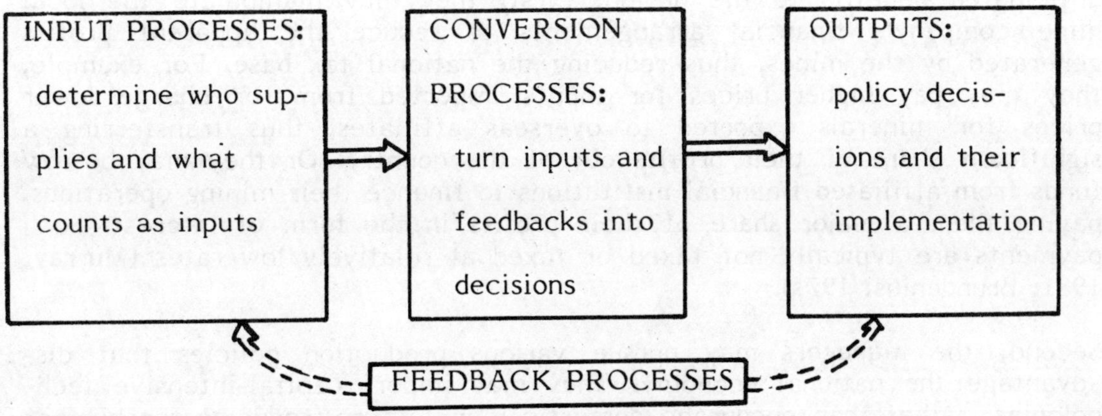

Note: Baratz and Bachrach (1970) criticised Dahl's original model (1961) as static. This model attempts to introduce more dynamic features by emphasizing, not particular inputs, feedbacks and outputs, but processes and the range of outputs (see R.B.Seidman, 1978).

The decision-making model in Figure 1 may help to understand the way the decision-making processes in large institutions function to influence policies.

In this brief talk, I will try to show that analysis of the major factors influencing key individuals behaviour within the relevant decision-making institutions helps to explain the newly independent governments relative lack of success in achieving their goals despite their efforts to implement several kinds of policy measures. These include attempts to: 1) capture the investable surplus directly; 2) ac quire shares of ownership to increase their share of profits and influence mine development; 3) create government mining corporations to implement proposed policies; and 4) establish mineral marketing boards.

In examining eachcategory, I will, first, make explicit the underlying, but frequently unstated premises; and then summarize the evidence as to the policy's impact on the inherited decision-making institutions in relation to the newly-formulated national goals.

a. Direct efforts to capture the investable surpluses:

As its underlying premise, this approach assumes as given the inherited constitutional framework that delegates mine ownership to private (frequently foreign) firms. It accepts, however, that governments may introduce laws to capture a greater share of the profits for the nation through various devices, including taxes, increased minimum wages, the introduction of development bonds, and more effective exchange controls. Several factors hindered the introduction of these kinds of laws. Among these, conventional wisdom, reinforced by international agencies like the World Bank and the International Monetary Fund, argued that such measures not only discouraged additional investment by existing firms, but alsofrightened away other potential investors (e.g. World Bank, 1981).

Moreover, this approach leaves the existing input, conversion and feedback processes intact in the hands of transnational corporate mining managers, whose careers primarily depend on maxmizing their parent companies global profits. They often respond to new laws in ways that limit the anticipated benefits to the nation: First, they may manipulate the local mine company's financial arrangements to reduce the apparent profits generated by the mines, thus reducing the national tax base. For example, they may pay higher prices for inputs imported from, or charge lower prices for minerals exported to overseas affiliates, thus transferring a significant share of their profits out of the country. Or they may borrow funds from affiliated financial institutions to finance their mining operations, paying them a major share of their profits in the form of interest. Such payments are typically not taxed or taxed at relatively low rates (Murray, 1981; Brundenius, 1975).

Second, the managers may pursue various production policies that disadvantage the national economy. They may import capital-intensive technologies, rather than encourage domestic manufacture, reducing the number of workers employed. They may mine more easily-reached valuable deposits to increase short-term profits after the taxes, leaving untouched marginal mineral reserves that, unless mined along with the richer deposits, will never be viable (Fifrench-Davis, 1975).

Third, transnational corporations may utilize their international marketing networks to reduce output in the country where taxes are high, while expanding it in those providing more favourable conditions. When Manley's government raised taxes on the bauxite mining firms in Jamaica, the transnational aluminium companies shifted their purchases to Australian and New Guinean mines, threatening Jamaïcan employment and incomes (Payne, 1984a). Where transnationals have been able toexpand output in new areas, or have developed substitutes, efforts to create producer associations to coordinate member countries output and prices to maximize national returns have failed (for the case of copper, see Takeuchi, 1975; Metzger,1975[3]

In short, governments' direct measures tocapture the mines investable surplus for national welfare have tended to fail because they left the transnational mining companies in control of the key operational decisions.

b. Government acquisition of shares:

Those advocating government acquisition of a share of the ownership of the national mines tended to assume this would automatically ensure government received anequivalent share of the mining profits. Government ownership of a majority of shares, furthermore, would enable it to direct mining operations along lines more beneficial to the nation.

Advocates of this position sometimes argue that the foreign firm has long drained profits more than equal its original investment, so the government may take over its local assets without paying for them (e.g. Chile under Allende). To avoid the danger of international sanctions, however,

(3) The most successful producer association, OPEC, now confronts tumbling prices in part because transnationals, with major assistance from their home country governments, have been able to expand output in non-OPEC member countries.

host governments more typically pay compensation, often in the form of a percentage of future profits. As Zambia discovered when it purchased a majority of the shares of its mines, however, this may consume a significant portion of the investable surplus the mine generates (Sharma, 1976).

Unfortunately, acquisition of shares, even a majority of shares, does not ensure the government will receive an equivalent share of profits, far less control mining and marketing operations. Evenin developed countries, it has long been recognized that eestablishment of boards of directors separates share owners from the corporate managers' daily decision-making processes (Berle and Means, 1933). In developing countries, where past company policies have denied nationals essential managerial skills, governments frequently negotiate management contracts with the transnational mining company (e.g. Passara, 1976; Ushewokunze, 1975). Transnational corporate managers thus continue to control the input, conversion and feedback processes that govern the local mining firms' policy decisions. In co-operation with remote head offices,they can easily manipulate price and profit decisions to reduce government's profit share. After Zambian government acquired 51 percent of the share of the two mining companies that dominate the national economy(4), for example, the cost of operating the mines unaccountably jumped 20 percent, reducing the government's profit share by an equivalent amount.

Furthermore, overworked government directors typically have neither the knowledge nor the opportunity to influence day-to-day decisions of the mining company's

---

(4) The transnational parent of one, the Anglo American Group, constituted the leading mining finance house in South Africa, with close ties to both British and U.S. financial and mining interests. Its investments spread throughout the independent countries of southern Africa. The other, a U.S. firm, American Metal Climax (AMAX), also had extensive investments, frequently in close collaboration with the Anglo American Group in Southern Africa, as well as the Americas (Seidman and Makgetla, 1980).

---

management(5). They must rely on the mine managers for the engineering data and financial information that determines their decisions as to which deposits to mine; the technologies to use and the prices paid for them; the numbers and wages of workers employed; whether or not and how much to borrow abroad, and the rates of interest paid; and even the amounts of minerals to market and where and a what prices to sell them. Moreover, the managers control the operative conversion processes relating to many day-to-day decisions that never reach the boards of directors. Typically, they also control the feedback processes. Few government representatives have the opportunity to talk directly to small local businesses that could have provided spare parts or components that managers ordered, instead,

---

(5) Sometimes one minister or permanent secretary will sit on 15 or more boards of companies in which the government holds shares (R.B.Seidman,1978).

from abroad. Nor do they have much occasion to hear personally from workers who object to management decisions converning wages and working conditions. Finally, the company managers frequently select and train the nationals who join the management team, and may convince them to adopt company, rather than national, perspectives and methods of operation.

It is small wonder that, in recent years, transnational corporate managers have even come to welcome third world governments' acquisition of shares in their local affiliates (e.g. silitshena, 1975). The government thus helps to finance the mines' operations, and will more likely favour company needs for foreign exchange or in dealing with mine workers demands for better wages and working conditions.

c. Government corporations:

To acquire greater control, some governments create state corporations to take over the management of the local mining company. They also, of course, confront the issue of whether and how much compensation to pay. Moreover, this attempt to establish greater state control has often turned out to be illusory. First, unless the government can hire qualified nationals, it may still have to negotiate management contracts that leave transnational corporate managers to determine day-to-day mining policies.

Second, the government often passes legislation that permits the state corporation to operate as much as possible like an autonomous private enterprise (United Nations, 1974)[(6)]. In third world mining economies, this may produce an anomalous situation in which the state-owned corporation almost constitutes a state-within-a-state. Its management not only makes decisions that affect the lives of thousands of workers and their families. Its policies control a major share of the nation's exports and incomes. It may exert a major influence on government policies. For a time in Indonesia, for example, the military officer heading the government petroleum corporation operated it entirely outside government control, amassing assets exceeding those of the state. The corporation funnelled funds to his supporters in the army. Some observers suggested that he, not the nominal head of state ruled the country (May, 1978; Crouch, 1978).

d. Government marketing boards:

In an effort to gaingreater control over marketing the mining sector's output, several governments have attempted to set up their own marketing agencies. In the mid '70s, Zambia's government established MEMACO (Metals Marketing Corporation) with a London Office to handle its mine exports. MEMACO appointed as its marketing agents several transnational corporations with which it had previously had relationships. These included

---

(6) In former British colonies, the drafters followed the British tradition based on the assumption that, unlike civil service bureaucratic red tape, private sector initiative in seeking to maximize profit stimulated efficiency (R.B.Seidman, 1978).

the South African mining finance house, Anglo American, which still owned minority shares in the Zambia mines; and two Japanese companies, Mitsui and Misubishi, which purchased a major share of the mines output (MEMACO annual reports).

Zambabwe's post-independence government -- aware of the high cost and difficulty of asserting national control through direct ownership, especially given the existence of manyrelatively small mines -- decided at the outset to create a state marketing board to handle mineral exports. Through the board, the Ministry of Mines sought to diversify the mines' export markets, selling minerals to companies and countries other than those linked to the operating transnational affiliates. In the process, the Ministry hoped to obtain a more accurate assessment of the value of the minerals sold, and thwart the loss of profits through transfer pricing. Over time, the Ministry anticipated the board would help tocapture a significant share of the investable surplus generated by the mines, as well as to train personnel to conduct research, manage, process and market minerals.

In a compromise designed to soften mining companay opposition to the board, however, the Zimbabwe government appointed as the marketing board's first manager a man, by the wonderful name of Mark Rule, who had played a prominent role in the pre-independence government's efforts to evade United Nation's sanctions. Although, in 1985, the board reported a profit of about a million Zimbabwe dollars(7), only further research could assess the extent to which the new board really altered the mining firms' decision-making processes.

In sum, newly independent governments' efforts to enhance the mining sector's contribution to national welfare failed, at least in part, because they did not fundamentally change the factors, particularly the decision-making structures, shaping the responsible managers' policies.

2. The need for a national development strategy:

In my opinion, a more important reason for the difficulties encountered by post-independence efforts to ensure the mining sector contributed more to national welfare lies in the governments failure adequately to address the fundamental issue of how to incorporate the mining sector in the overall development process. Most focused only on parts of the problem: How to capture a greater share of the investable surpluses generated by the mines; or how to assert greater government control over the mines or mineral marketing to increase their sales to earn more foreign exchange and revenue. Many spent much of the investable surplus they did capture to finance expanded social services and economic infrastructure, rather than investing in productive activity that, over time, would generate greater income and employment. Implicity, they adopted the western orthodoxy that investment in the mines and associated infrastructure would initiate an automatic multiplier effect, spreading increasingly productive employment activities throughout the rest of the economy. They neglected investments in a whole range of possible productive spin-off activities that could have stimulated more balanced, integrated national

(7) A tiny fraction (less than 0.2 percent) of the value of the nation's mineral export (Zimbabwe Statistical Abstract, Oct. 1985)

development.

To illustrate: In Zambia, the government seemed to view the mining sector as a goose that laid golden eggs. All it had to do was collect them to finance its 'humanist' social welfare programs. In 1969, when the government took over 51 percent of the mining companies, the President explained that it primarily aimed to ensure that they invested to expand the production and sale of copper[8]. In its agreement with the companies, the government required neither an affirmative action program to train local managers, nor investment in more advanced stages of processing. Instead, it went so far as to grant the companies permission to remit, tax free, one kwacha for every kwacha invested in expanding copper output. The companies simply borrowed money abroad to invest in the mines (leaving the government, as a majority-share owner, responsible for repaying the debt before receiving any profits), and remitted, tax free, their share of the profits to their parent companies. Government revenues from the copper mines plummed from K235 million in 1969 to K55 million in 1972 (Sharma,1975).

Far from contributing to increased national prosperity, the resulting expansion of Zambian copper sales -- along with that of other third world countries pursuing parallel policies -- predictably contributed to an oversupply that pushed world copper prices down in a semi-permanent slump. Simply to maintain its level of import-dependent development, Zambia's government like those of other mineral-exporting countries, had to borrow heavily abroad. It turned to the IMF for help. By the beginning of the '80s, Zambia had received about a third of the resources the IMF made available in all of the southern Africa -- and, with them, the typical IMF austerity package (Seidman, 1985; Makgetla, forthcoming). Under IMF pressure, the government re-opened the economy to South African imports, undermining domestic import-substitution industrial production and

---

(8) The firms had been remitting up to 90 percent of their after-tax profits abroad.

---

employment[9]. The economy stagnated. Unemployment mounted. Rising prices slashed real incomes (Makgetla, 1985; Mudenda, 1985).

The governments of Zambia and other mining economies could have learned from many years of experience that miing projects in third world countries, unlike those established earlier in Europe and the United States, seldom

---

(9) After independence, the government had encouraged import-substitution industrialization, acquiring shares in manufacturing firms, and providing incentives for managers to expand domestic production. As in the case of import substitution elsewhere, however, the industries mainly located in already-developed areas along the line-of-rail, produced luxuries and semi-luxuries, and depended on imported capital intensive machinery and equipment that provided relatively little employment. Essentially, in other words, they tended to aggravate the nation's dualistic, external dependence (Turok, 1980).

automaticallystimulate the spread of development. The introduction of the Tata steel industry in India, for example, had a different impact from that established in the 19th century in the United States. The U.S. steel industry grew in response to the expanding domestic demand created first by the railroads, and later by the auto and construction industries. In contrast, extensive investigation has revealed that for decades the Tata steel industry contributed relatively little to increased productivity and rising living standards in the surrounding Jamshedpur region. Relatively self-sufficient neighbouring villages provided no market for steel or steel-based products. Local blacksmiths lacked the capital and skills toset in motion the development process, generating a demand for steel. Moreover, as the steel industry aged, new capital 'modernization' expenditures tended to limit rather than expand employment. Only after World War II, by providing additional employment and creating a steadilyexpanding demand for component parts as well as consumer goods and services, did government-assisted expansion of steel-using commercial vehicle and machinery manufacture stimulate regional growth (Gupta, 1983).

The complexity and large scale of modern technologies associated with mining today thwart their potential as poles of growth that will automatically foster integrated development throughout third world regions[10]. The Jamshedpur experience underscores the fact that, today, third world efforts to take advantage of mining potentials tospur regional development have become "Dincreasingly dependent on social action, policy and organization." (Gupta, 1983, 66).

This is borne out by the South African case. There, the racist regime aggressively exercised state power to spur the multiplication of the transnational corporate manufacturing investments that transformed that country from a predominantly mining economy to an industrialized one, now characterized by the international agencies as "developed" (e.g. United Nations Statistical Yearbook). The apartheid state deliberately coerced the black majority into impoverished reserves to provide cheap labour for the minority-owned mines, commercial farms and factories. But we might ponder the implications of the fact that, over the last 25 years, the regime played a crucial role in directing investments, not only by the domestic mining finance houses, but also by transnational corporations from all the major western nations, to the construction of its modern military-industrial complex. Thus the regime strengthened South Africa's capacity as a regional subcenter, enabling it to marginalize and exploit, not only its own population, but the peoples of the neighbouring states (Seidman and Makgetla,1980). To reduce dependence on black labour after the 1976 Soweto uprising, the regime deliberately fostered the capital-intensive mechanization of mines and agriculture that today has left half the black population without land or jobs, starving in fragmented scraps of the least fertile, least well-watered 13 percent of the nation's land: the "bantustans" (A.Seidman, 1985). After winning liberation, the people of South Africa will need to study these developments, as well as learn from other countries experiences, to create a democratic state and restructure the institutions controlling

---

(10) For a definition and illustration of the concept of growth poles, see Green and Seidman, 1968. Mabogunje, (1978) reviews of the theory of growth poles in third world economies, and calls for further research concerning their potential role in Africa. Gupta (1983) discusses growth pole theory in relation to the Tata iron and steel industry.

the mines and other productive activities, not to further enrich a minority, but to meet their basic needs.

In short, the available evidence underscores the necessity for developing country governments to formulate longterm plans, both for improving the pattern of resource allocation which mining investment may make possible: and for introducing the essential changes in inherited sets of key institutions to implement those plans.

a. Longterm plans for resource development:

Governments should plan their longterm development strategies to incorporate all the potential linkages and spinoffs from the mining sector (Tironi, 1975; A.Seidman, 1975b). These include:

*The production of building materials and consumer goods, including food for the labour force, during the construction of the mine works and associated infrastructure: Integrated into on-going plans for development of the mining project, these activities should contribute to the spread of permanent employment opportunities in agriculture and small scale rural industries.

*The processing of the mine's output, not only to increase its export value, but also to facilitate its use in the fôrm of inputs into other sectors of the national and regional economy: Iron processed into steel may be used for all kinds of building materials and manufactured products; copper may be used for electric wires and appliances, as well as in brass; aluminium made from bauxite provide a durable light construction material for many purposes; etc. Planning agencies should make longterm plans, not only for the physical location and technology of these industries, but also to provide adequate manpower and finance to ensure their realization.

*The construction infrastructure associated with the mining project to serve a double purpose of fulfilling the development needs (including productive activities) of the neighbouring communities: Rather than focusing solely on the mine project's needs in selecting and building a particular type of infrastructural projects -- roads, railroads, bridges, water supply, power, and so forth -- planners should also design it to maximize their potential stimulation of agriculture and small scale industries.

*The training and involvement of the workers at every stage and every level to improve their incentives and efficiency in increasing productivity: In particular, nationals should learn engineering, accounting, marketing and the other skills needed to manage every aspect of the operation of the mines and associated infrastructure in relation to national welfare. The semi- and unskilled workers on the mines and associated infrastructural projects, too, should receive regular on-the-job training and special courses. This should enable them, not only to participate more fully in the efficient operation of the mines, but also to contribute to technological innovation if they later work in other sectors of the national economy.

*Development of local research and teaching in national universities and specialized mining institutes like the Indian School of Mines to

ensure
/the national economy realizes the full growth pole potential of miing projects. Research should not focus only on the most appropriate technologies for the mineral's extraction, processing, and linkages to the rest of the regional and national economy. It should investigate broader issues like the conditions of the national, regional and world markets and the appropriate incentives and conditions to spur increased worker productivity. Wherever possible, research should encourage participation by the miners and workers in neighbouring communities tostimulate the creative improvement of their performance and the quality of the goods they produce.

today, given the economies of scale characteristic of large mining and associated processing industries, small mining economies could more easily plan these kinds of linkages if they could cooperate with their neighbours. The former colonial rulers carved up entire regions into mini-economies without regard to ethnic, geographical or economic realities, leaving the succeeding independent nations too small to provide the markets or resources needed to create large-scale modern industrial development based on their mineral resources(11). Separately, little mining economies like Zambia's or Jamaica's with low per capita incomes and populations smaller than those of a good size European city, cannot easily escape dependence on external markets, technologies, and sources of finance dominated by trans-national firms. This urges research into the possibilities of incorporating their national mining sectors into regional strategies for industrialization.(12) Several models of regional cooperation exist.

b. Planning for essential institutional changes:

The experience of developing countries the world around emphasizes that, to realize the planned development of mining projects as poles of growth, govern -ments must change the inherited institutions that have shaped their past disarticulated development. Given the constraints on their available manpower and finance, however, they cannot change all of them at once. They need to identify priorities. It only makes sense that they should start with those institutions that control their nations 'commanding heightss', that is, those with the greatest linkages and influence throughout the economy. In particular, this implies asserting control over basic industries; export-import and internal wholesale trade; and financial institutions (Nyerere, 1967; UN 1974).

The large potential gains from integrating mining sectors into regional industrialization strategies suggest, furthermore, the necessity of considering, not only how toalter the institutions to increase control over the national commanding heights, but simultaneously, how to restructure them to facilitate regional cooperation. Evidence shows that formation of common markets that leave investment decisions to private profit maximizing enterprise (domestic or foreign), tends to reinforce and aggravate uneven development and perpetuate external dependence (Green and Seidman, 1968; Echeveria ~~et al,~~ 1981). Moreover, newly liberated peoples typically cherish the national

---

(11). This is true throughout many parts of the world (Green and Seidman, 1968; Salgado,1978; Pxline,1977; Belassa and Stroutjesdijk,1977; Gupta,1983; Ballance et al,1982). Even larger countries, like Nigeria, Brazil and India, remain constitutionally divided into states whose separate administrative structures tend to hinder formulation and implementation of national development strategies (Kuklinksy, 1978).

(12). In the developed world, these include the European Economic Community (EEC- Demco, 1984; Franzmeyer, 1983; Hoffman, 1984) and the Community for Mutual Economic Assistance (CMEA - Saunders, 1983; Hoffman, 1984) and, among developing countries, Caricom (Harrison, 1967),. the Andes Pact (Echeveria et at, 1981), the Association of South East Asian Nations (ASEAN - Castro, 1982; Ghosh, 1984), the Southern African Development Coordination Conference (SADCC - Thompson, 1985) and the Economic Community for West Africa (ECOWAS - Asante,1984).

---

identity they have won through prolonged struggle. Their national governmental institutions may focus on realizing national potentials, implementing policies that may conflict with attainment of regional possibilities[13]. This suggests that, while examining the national economies commanding heights, researchers should also explore institutional changes to achieve the minimal threshhold of cooperation required to implement an effective regional strategy.

The mines and associated industries, because of their key role in producing a major share of a third world mining economy's exports, incomes and investable surpluses, as well as their potential for stimulating the spread of productivity in all sectors of the economy, constitute one of any mining economy's "commanding heights." The evidence I have given relating to my first proposition, however, indicated that, in many if not most third world countries, the institutions to which post-independence governments delegated the power to manage the mines continued to grind out decisions much like those they made in the past. Changes in the formal ownership of the mines did not sufficiently alter those decision-making processes to produce different results. Furthermore, the small size of many mining economies may render it difficult, if not impossible, for them, alone, to exert significant control over the international marketing networks dominated by transnational corporations. Unless they can link their mining sector into a regional industrialization strategy, their best option may be to treat it as a money-spinner, capturing and investing a major share of the surplus in a more modest national industrialization program to meet the peoples' needs. Self-evidently, this will work only where the mining venture is sufficiently profitable to attract a foreign partner while providing the government with an adequate share.

Unless a government successfully alters the inherited processes determining the inputs, conversion and feedback in the mining companies, however, the mining sector and associated industries will not likely conform with plans designed to implement either a regionally or a nationally-oriented development strategy. National research and teaching centers like the Indian School of Mines could make a valuable contribution by investigating the kinds of institutional changes that, in particular countries, would

---

(13). This has been illustrated the world around (Mazzeo, 1984; Axline, 1977; Ballance et all, 1982; Belassa and Stroutjesdijk, 1978; Castro,1982)

help to ensure these industries played a more positive role in national and regional development.

In the typical third world mining economy which relies on imported technologies and, to remain viable, must export at least some of its mineral output, foreign trade constitutes a second "commanding height". Examination of the decision-making structures characteristic of most nations trading institutions, however, reveals features essentially similar to those of the mining firms. They often delegate control over this sector to private firms, frequently transnational corporate affiliates the mining transnationals (of. Cobbe,1979). These typically decide not only whether, where and at what prices to market the minerals and, if any, the manufactures produced, but also what technologies and other inputs to import for the mines and all other sectors. Many third world governments establish marketing boards kfor major agricultural exports, but Zimbabwe is one of the few governments that has set up such a board for minerals. Most usually exercise indirect regulation, of exports, more or less in line with perceived national interests, through licensing, foreign exchange and tariffs. In some countries, state trading companies compete with private importing firms, and in a relatively few cases have largelyr eplaced them. In each national and regional a context, researchers should explore the particular constraints and resources as a foundation for designing more appropriate trading institutions and policies.

Financial institutions, which control the accumulation and reinvestment of a nation's investable surpluses, constitute the third economic "commanding height". Financial institutions help determine how much of the surplus generated by the mines and other sectors is remitted abroad; and whether the share that remains is reinvested primarily to expand the mining sector, or in more balanced, integrated national or regional development. At independence, most developing country governments only intervened directly in the financial arena through the central bank and by taxing corporate incomes. They typically dealegated the major decision-making role to private, often foreign, financial institutions: commercial banks, pension and insurance funds, and various kinds of investment banks. Even when governments acquired shares of ownership in some of these institutions, they frequently left key decisions to managers whose ties or predilections shaped decisions favourable to private rather than national interests (A. Seidman, 1986a). Investigation of this third crucial area, should help discover the kinds of institutional changes required in any particular country to ensure re-allocation of investable surpluses to more self-reliant, balanced development capable of ensuring increasingly productive employment and rising living standards for all the nation's, or, better yet, all the region's inhabitants.

CONCLUSION:

In this brief talk, I have not attempted to shed new light on the role of the mining industry in third world development. Instead, I have tried to identify some of the lessons of past experience, and outline future research which could contribute to planning and implementing more effective strategies to incorporate mining into overall development. I have suggested that the relative failure of many governments' to enhance their mining sectors' contribution to national welfare lies only in part inthe inadequacy of the changes they made in the inherited decision-making structures governing the mines' operation. Equally, if not more important, they seldom integrated the mines as poles of growth into long term plans to contribute to restructuring either their national or the regional economies.

Not that political leaders and government officials have not called for sweeping changes in the dualistic externally dependent allocation of resources, including the mining sectors, that condemned their populations to poverty. Many have also united to call for regional cooperation. But they seldom implemented the necessary step-by-step alterations in the inherited sets of institutions governing their nations' economic commanding heights' to carry out the proposed reallocationof resources at either the national or regional level.

If this hypothesis is valid, it poses a challenge for third world teaching and research centers like the Indian School of Mines. Their faculties and students can pioneer in investigating the practical step-by-step changes, not only in resource allocation, but also in key institutions to ensure that existing or proposed mining operations contribute to all aspects of national, and, in so far as possible, regional development. By participating in this kind of research, furthermore, the students in these centers will acquire the training and perspectives necessary to provide leadership in formulating and implementing plans to ensure that mineral wealth contributes to the spread of increasingly productive employment opportunities and rising living standards for all the inhabitants of their nation, and perhaps even for the region.